Katharina L. Dürr

Functional Significance of Na,K- and H,K-ATPase β-Subunits

Katharina L. Dürr

Functional Significance of Na,K- and H,K-ATPase β-Subunits

studied by Voltage-Clamp Fluorometry in Xenopus Oocytes

Südwestdeutscher Verlag für Hochschulschriften

Impressum/Imprint (nur für Deutschland/ only for Germany)
Bibliografische Information der Deutschen Nationalbibliothek: Die Deutsche Nationalbibliothek verzeichnet diese Publikation in der Deutschen Nationalbibliografie; detaillierte bibliografische Daten sind im Internet über http://dnb.d-nb.de abrufbar.

Alle in diesem Buch genannten Marken und Produktnamen unterliegen warenzeichen-, marken- oder patentrechtlichem Schutz bzw. sind Warenzeichen oder eingetragene Warenzeichen der jeweiligen Inhaber. Die Wiedergabe von Marken, Produktnamen, Gebrauchsnamen, Handelsnamen, Warenbezeichnungen u.s.w. in diesem Werk berechtigt auch ohne besondere Kennzeichnung nicht zu der Annahme, dass solche Namen im Sinne der Warenzeichen- und Markenschutzgesetzgebung als frei zu betrachten wären und daher von jedermann benutzt werden dürften.

Verlag: Südwestdeutscher Verlag für Hochschulschriften Aktiengesellschaft & Co. KG
Dudweiler Landstr. 99, 66123 Saarbrücken, Deutschland
Telefon +49 681 37 20 271-1, Telefax +49 681 37 20 271-0
Email: info@svh-verlag.de
Zugl.: Berlin, TU, Diss., 2009

Herstellung in Deutschland:
Schaltungsdienst Lange o.H.G., Berlin
Books on Demand GmbH, Norderstedt
Reha GmbH, Saarbrücken
Amazon Distribution GmbH, Leipzig
ISBN: 978-3-8381-1349-4

Imprint (only for USA, GB)
Bibliographic information published by the Deutsche Nationalbibliothek: The Deutsche Nationalbibliothek lists this publication in the Deutsche Nationalbibliografie; detailed bibliographic data are available in the Internet at http://dnb.d-nb.de.

Any brand names and product names mentioned in this book are subject to trademark, brand or patent protection and are trademarks or registered trademarks of their respective holders. The use of brand names, product names, common names, trade names, product descriptions etc. even without a particular marking in this works is in no way to be construed to mean that such names may be regarded as unrestricted in respect of trademark and brand protection legislation and could thus be used by anyone.

Publisher: Südwestdeutscher Verlag für Hochschulschriften Aktiengesellschaft & Co. KG
Dudweiler Landstr. 99, 66123 Saarbrücken, Germany
Phone +49 681 37 20 271-1, Fax +49 681 37 20 271-0
Email: info@svh-verlag.de

Printed in the U.S.A.
Printed in the U.K. by (see last page)
ISBN: 978-3-8381-1349-4

Copyright © 2010 by the author and Südwestdeutscher Verlag für Hochschulschriften Aktiengesellschaft & Co. KG and licensors
All rights reserved. Saarbrücken 2010

Contents

1. **General Introduction**   5
   1.1. P-Type ATPases 6
      1.1.1. The Catalytic Cycle of P-Type ATPases 7
      1.1.2. Electrogenicity of P-Type ATPases 8
      1.1.3. Structure of P-Type ATPases 11
      1.1.4. Oligomeric P-Type ATPases 14
   1.2. Physiological Importance of ATPases 17
      1.2.1. Physiological Functions of the Na,K-ATPase 17
         1.2.1.1. Na,K-ATPase Inhibitors 18
      1.2.2. Physiological Functions of H,K-ATPases 20
         1.2.2.1. The nongastric H,K-ATPase 20
         1.2.2.2. Localization and Regulation of gastric H,K-ATPase 20
         1.2.2.3. Inhibitors of gastric H,K-ATPase 24

2. **Material and Methods**   31
   2.1. Molecular biology 32
      2.1.1. Expression vectors and cDNA-constructs 32
      2.1.2. Site-directed mutagenesis 33
   2.2. Oocyte preparation and cRNA injection 33
   2.3. Membrane preparation from *Xenopus laevis* oocytes 33
      2.3.1. Isolation of plasma membranes 33
      2.3.2. Preparation of total membranes from *Xenopus laevis* oocytes 34
      2.3.3. Immunoblotting (Western Blot) 34
   2.4. Rb^+ uptake measurements 34
      2.4.1. Rb^+ uptake measurements 34
      2.4.2. Atomic Absorption spectroscopy (AAS) 34
   2.5. Electrophysiology 35
      2.5.1. Oocyte pretreatment, fluorescence labeling and experimental solutions 35
      2.5.2. Voltage-clamp fluorometry 35
      2.5.3. Analysis of stationary currents of the Na,K-ATPase 35
      2.5.4. Analysis of transient currents of the Na,K-ATPase 35
   2.6. Extracellular pH measurements 36

3. **Properties of glycosylation-deficient Na,K- and H,K-ATPase**   37
   3.1. Introduction 38
   3.2. Results 40
      3.2.1. Plasma Membrane Delivery and α-Subunit Stabilization of Glycosylation-Deficient Mutants 40
      3.2.2. Enzyme Activity of Glycosylation-Deficient Mutants 41
      3.2.3. E_1P/E_2P Conformational Distribution and Kinetics of $E_1P{\rightarrow}E_2P$ Transition of Glycosylation-Deficient Mutants 42
   3.3. Discussion 44

Contents

4. E_2-stabilizing α-TM7/β-TM interactions of Na,K- and H,K-ATPase   **49**
   4.1. Introduction 50
   4.2. Results & Discussion 52
       4.2.1. E_1P/E_2P conformational distribution and kinetics of the E_1P/E_2P-transition for Na,K-ATPase wildtype and β−(Y39W,Y43W) mutant enzymes 52
       4.2.2. Molecular determinants for the E_2-stabilizing effect mediated by the two conserved tyrosines of the β-subunit 54
       4.2.3. E_1P/E_2P-conformational distribution, apparent Rb^+ affinities and SCH 28080 sensitivity of the H,K-ATPase wildtype and β-(Y44W,Y48W) variant enzymes   55
       4.2.4. H^+ secretion of H,K-ATPase wildtype and β-(Y44W,Y48W) variant enzymes . 58
       4.2.5. TM7 residues of Na,K- and H,K-ATPase α-subunits relevant for the E_2-stabilizing interaction with the two conserved β-tyrosines 60
       4.2.6. Physiological relevance of the β-subunits' E_2-stabilizing effect for Na,K- and gastric H,K-ATPase activity *in situ* 65

5. Functional Significance of the β-N-terminus of gastric H,K-ATPase   **67**
   5.1. Introduction 68
   5.2. Results 68
       5.2.1. Cell surface expression of N-terminally deleted H,K β-variants. 68
       5.2.2. E_1P/E_2P conformational distribution of N-terminally deleted H,K-ATPase β-mutants. 69
       5.2.3. Rb^+uptake kinetics and SCH 28080 sensitivity of N-terminally truncated H,K-ATPase β-subunits 72
   5.3. Discussion 74
       5.3.1. Physiological significance of the β-N-terminus for gastric H,K-ATPase activity *in situ* 74
       5.3.2. Adverse conformational effects exerted by two contact sites between α-subunit and the β-N-terminus 74
       5.3.3. Comparison to the functional effects of β-N-terminally truncated Na,K-ATPase   75

6. Summary and Conclusions   **79**

A. Supplementary data   **85**

List of Figures   **91**

List of Tables   **93**

Bibliography   **95**

Abbreviations

AMOG	Adhesion Molecule On Glial Cells
APA	Acid Pump Antagonists
ECL	cells enterochromaffin-like cells
EM	Electron Microscopy
CHIF	Channel Inducing Factor (Corticosteroid Induced Factor)
COS	cell line, COS is an acronym, derived from the cells being CV-1 (simian) in Origin, and carrying the SV40 genetic material
FHM2	Familal Hemiplegic Migraine type 2
GERD	Gastro-esophageal reflux disease
HEK293	Human embryonic kidney cell line
HOMG2	Hypomagnesemia 2
LLC-PK	clonal pig kidney epithelial cell line
Mat-8	Mammary tumor marker 8
MDCK	Madin Darby canine kidney cell line
MFS	Major Facilitator Superfamily
NCX	Sodium Calcium exchanger
NSS	Neurotransmitter Sodium Symporter Family
PC	Phosphatidylcholin (Lecithin)
PCMA	Plasmamembrane Calcium ATPase
PE	Phosphatidylethanolamin (Cephalin)
PKA	Proteinkinase A
PLB	Phospholamban
PLM	Phospholemman
PPI	Proton Pump Inhibitors
PS	Phosphatidylserin
PUD	peptic ulcer disease
RDP	Rapid-onset Dystonia Parkinsonism
Ric	Related to ion channel
SERCA	Sarcoplasmatic Reticulum Calcium ATPase
SF9	insect cell line derived from the ovary tissue of *Spodoptera frugiperda*
TM	transmembrane domain
TMRM	Tetramethylrhodamine-6-maleimide
VCF	Voltage-Clamp Fluorometry

CHAPTER 1

General Introduction

1.1. P-Type ATPases

P-type ATPases make up a huge superfamily of ion pumps that use the energy of ATP hydrolysis to fuel the transmembrane transport of charged substrates across biological membranes. The "P" stands for phosporylation, since a hallmark of the more than 300 members of this family is the reversible formation of an acyl-intermediate in which the γ-phosphate of the enzymatically hydrolysed ATP is covalently linked to a highly conserved Asp side-chain. This Asp residue is part of the DKTGTLT-sequence motif which is located in the so-called "P-domain" and defines the membership of the P-type ATPase family. Two additional conserved motifs related to the phosphorylation and subsequent dephosphorylation reaction are the GDGXNDXP motif, present immediately after the ATP binding domain (N-domain, for "nucleotide-binding domain") and the TGES sequence located to the so-called actuator-domain (A-domain). P-type ATPase-encoding genes are found throughout all five kingdoms of life, but they are more widespread in eukaryotes than in bacteria and archea. Yet, the simplest and presumably most ancient ion pumps (type-I) are mainly found in bacteria, e.g. the Kdp K^+ pump from *E.coli*.

Based on sequence homology, the P-type ATPase family can be divided into 5 branches, referred to as types I-V (Axelsen and Palmgren, 1998) (see Table 1.1). Although the number of transmembrane domains varies between 6 to 10, all P-type ATPases have an even number of transmembrane domains with the N- and C-termini and the large catalytic (ATP-binding and -hydrolyzing) domain facing the cytoplasmic side of the membrane. Table 1.1 gives a brief summary of the various substrates that are transported by different members. Most of them translocate small, "hard" cations (H^+, Na^+, K^+, Ca^{2+} or Mg^{2+}), but some are transporting "soft" transition metal ions instead, including Cu^+, Ag^+, Zn^+, Cd^+, Hg^+ and Pb^+. Since these latter are heavy metal ions, the type-IB transporters have the physiologically important function of removing these toxic ions from the cell. Mutations in the human Cu^+ efflux pumps therefore cause the rare but lethal hereditary Menkes' and Wilson's diseases (Bull et al., 1993; Tanzi et al., 1993). The most unconventional substrates are transported by type-IV ATPases that are uniquely found in eukaryotes. They catalyze the movement of the aminophospholipids PS, PE or PC from the extracellular to the cytoplasmic leaflet of the plasma membrane lipid bilayer (Poulsen et al., 2008*b*). Since even these rather hydrophobic substrates all contain positively charged head groups, the cationic nature seems to be the common denominator of all P-type ATPase substrates. However, it should be noted that it is not proven yet that type-IV ATPases are actually translocating the cationic phospholipids themselves - it has been proposed instead that by transporting other cations they may possibly generate a concentration gradient that subsequently drives phospholipid translocation through another unidentified symporter (Kühlbrandt, 2004). Furthermore, the substrates that are transported by the type-V ATPases still remain to be identified.

Type	Gene	Name	Iso-forms	Organism	Cation specificity	TM	accessory subunits
I-A	kdpB	Kdp ATPase	1	E. coli	K^+	6 (or 7)	KdpA, KdpC, KdpF
I-B	copA	CopA	1	E. coli	$Cu^+\ Ag^+$	8	-
	zntA	ZntA	1	E. coli	$Zn^+,\ Pb^+$ $Cd^+,\ Hg^+$	8	-
	cadA	CadA	1	S. aureus	Cd^+	8	-
	ATP7A	MDAP	1	H. sapiens	Cu^{2+}	8	-
	ATP7B	WDAP	1	H. sapiens	Cu^{2+}	8	-
II-A	ATP2A	SERCA	3	H. sapiens	Ca^{2+}	10	phospholamban
II-B	ATP2B	PMCA	4	H. sapiens	Ca^{2+}	10	-
II-C	ATP1A (1-4)	Na,K-ATPase	4	H. sapiens	Na^+/K^+	10	β-subunit γ-subunit[1]
	ATP4A	gastric H,K-ATPase	1	H. sapiens	H^+/K^+	10	β-subunit
	ATP12A	non-gastric H,K-ATPase	1	H. sapiens	H^+/K^+ Na^+/K^+ NH_4^+/K^+	10	β-subunit
II-D	CA1	Ca^{2+} motive P-type ATPase	1	L. donovani	$Na^+,\ Ca^{2+}$	10	
III-A	AHA1-11	plant H-ATPase	11	A. thaliana	H^+	10	-
III-B	mgtA	Mg-ATPase	1	E. coli	Mg^{2+}	10	-
IV	APLT1	APLT1	1	L. donovani	PS	10	cdc50
	ATP8A1	APLT	14	H. sapiens	PS	10	cdc50
	Drs2	DRS2	1	S. cerevisiae	PS, PE	10	cdc50p
	Dnf1p	DNF1	3	S. cerevisiae	PS, PE, PC	10	Lem3p
V-A	At5g23630	Probable cation-transporting ATPase	1	A. thaliana	unknown	10	?
V-B	ATP13A (1-4)	Probable cation-transporting ATPase	4	H. sapiens	unknown	10	?

Table 1.1.: **Classification of P-Type ATPases**. For each of the five classes, the most prominent members (preferably human, if known) are listed.

1.1.1. The Catalytic Cycle of P-Type ATPases

Ion transport of all P-type ATPases is probably mediated by a common mechanism, where the reversible phosporylation of the conserved Asp plays a key role by facilitating the transition between two principal conformations with inwardly- and outwardly-facing binding sites for the respective transported cations. Consequently, P-type ATPases are also referred as "E_1/E_2-ATPases". It is assumed that the conformational transition between these two states is linked to changes in the respective affinities for the translocated cations, thus enabling uptake of the ions from one side and their subsequent release to the other side of the membrane. For P-type ATPases that transport two different ion species, the transport is strictly consecutive and occurs in opposite directions across the membrane ("ping-pong" or "alternating access" model). Between their binding and release at different sides of the membrane, the ions undergo "occluded states", in which the ions are trapped within the enzyme without access to either side of the membrane.

---
[1]or other members of the FXYD-family (see Table 1.2 on page 16)

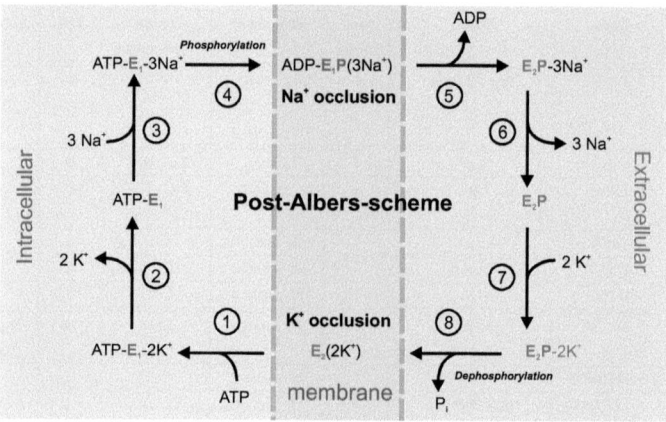

Figure 1.1.: **Post-Albers reaction cycle of the Na,K-ATPase.**

For the Na,K-ATPase, a detailed reaction scheme that accounts for these properties has been formulated (Albers, 1967; Post et al., 1972), the so-called "Post-Albers-Cycle" (see Figure 1.1). Starting from the intracellular side of the membrane, low-affinity binding of cytosolic ATP in the E_2-state facilitates the transition to the E_1-state (step 1 in Figure 1.1). The conformational change is accompanied with a reduction of the K^+ affinity, thus releasing the two bound K^+ ions to the cytoplasmic space (step 2). The E_2/E_1 transition also converts the ATP binding site into a "high affinity binding site". Furthermore, as a consequence of the increased Na^+ affinity in the E_1-state, three Na^+ ions are bound intracellularly (step 3), followed by phosphorylation of Asp369 (pig renal α_1-isoform numbering) by ATP (step 4), yielding E_1P with occluded Na^+ ions. The phosphorylation of Asp369 favours a conformational change to the E_2P-state (step 5). ADP dissociates from the N-Domain, Na^+ ions are deoccluded and released (step 6). Two extracellular K^+ ions bind to the high affinity binding sites (step 7) which in turn accelerates the dephosphorylation reaction (step 8), resulting in a K^+ occluded E_2-state of the enzyme.

Ion transport of other P-type ATPases most likely takes place according to similar reaction schemes, since the domains involved in nucleotide binding and phosphorylation are highly conserved.

1.1.2. Electrogenicity of P-Type ATPases

Since the substrates translocated by P-type ATPase are charged, their overall transport activity is electrogenic when one or more net charges (according to their respective stoichiometry) are moved across the membrane during a full reaction cycle. Famous examples are the Na,K-ATPase with a 3 $Na^+/2~K^+/1$ ATP stoichiometry, which moves one net charge per reaction cycle and the SERCA with a 2 $Ca^{2+}/2~H^+/1$ ATP stoichiometry, thus moving two net charges per pump cycle. As a consequence of these net charge movements, the pumps act as generators of electric current, thereby hypercharging the membrane capacitance which can be monitored by fluorescent styryl dyes of the RH family, such as RH-160, RH-237 or RH-421 (Klodos and Forbush, 1988). Due to their amphiphilic character, these compounds are inserted into lipid bilayers where their fluorescence is influenced by changes in the local electric field (electrochromism).

Electrogenic steps of the H,K-ATPase reaction cycle

As a consequence of their charge-transporting activity, the turnover number of electrogenic P-type ATPases may depend on the membrane potential. This happens if either a rate-limiting step itself or a step controlling the level of an enzyme intermediate entering a rate-limiting step is electrogenic. Therefore, the turnover number of an overall electroneutrally operating ATPase may be potential-sensitive, provided that the rate-limiting step is accompanied by net movement of charge across the transmembrane electric field. For example, the rate of H^+ secretion by the electroneutral gastric H,K-ATPase (pump stoichiometry of 2 H^+/ 2 K^+/ 1 ATP) was shown to be voltage-dependent *in situ* (Rehm, 1965). Since positive membrane potentials enhance the pump rate of the enzyme, an electrogenic step in the H^+ outward transporting limb appears to be rate-limiting under these conditions. Additional evidence for the presence of an electrogenic step in the H^+ limb was provided by black lipid membrane measurements in which H,K-ATPase containing parietal cell membrane fragments were adsorbed to a planar lipid membrane (van der Hijden et al., 1990; Stengelin et al., 1993). Upon light-induced release of caged ATP, a transient current was observed that increased with higher ATP concentrations but decreased and eventually disappeared with increasing K^+ concentrations. The observed transient charge movement was therefore assigned to the proton movement during the H^+-translocating branch of the catalytic cycle.

Interestingly, K^+ inhibition experiments on inside-out gastric vesicles revealed that another electrogenic step exists in the K^+-branch of the H,K-ATPase reaction cycle (Lorentzon et al., 1988). These studies showed that the inhibitory effect of high intracellular K^+ concentrations on ATPase activity is potential-sensitive: K^+ inhibition was efficiently prevented at interior positive diffusion potentials in presence of the K^+-specific ionophore valinomycin (corresponding to negative intracellular potentials), but could be restored by adding a protonophore that dissipated the K^+ diffusion potential.

Electrogenic steps of the Na,K-ATPase reaction cycle

Numerous studies have revealed that also several steps in the sodium pump reaction cycle are electrogenic. Information about the electrogenic steps can be obtained from the voltage-dependence of steady-state pump currents under various ionic conditions where different steps become rate-limiting for the overall process. For example, in Na^+-containing extracellular solutions with saturating concentrations of K^+, the pump currents decrease at hyperpolarizing potentials, which can be attributed to enhanced electrogenic reverse binding of extracellular Na^+. This competes with binding of extracellular K^+ ions, thus lowering the forward pumping rate. In contrast, the I-V relationship is relatively voltage-independent in the absence of extracellular Na^+ at saturating K^+. Below K^+ saturation in extracellular Na^+-free solutions, the pump currents are even slightly increased at hyperpolarizing potentials (Rakowski et al., 1991), since the effective concentration of K^+ at the external binding sites is increased by the applied voltage, resulting in enhanced turnover. This indicates that also extracellular K^+binding is electrogenic, albeit more weakly than extracellular Na^+ binding.

At low intracellular Na^+ concentrations and saturating extracellular potassium levels in Na^+-free solutions, the weak electrogenic binding of sodium ions to the intracellular binding site is revealed. If under these conditions, the intracellular Na^+ concentration is increased to saturation, the voltage-dependent pump currents still exhibit a shallow positive slope (Rakowski et al., 1997), thus indicating that also the inward-to-outward-facing conformational change ($E_1P \rightarrow E_2P$) is slightly electrogenic and rate-limiting for Na^+ forward transport. Therefore, at least the voltage-dependence of this step is not compatible with the commonly used access channel model (see below).

Another possibility to study the electrogenic steps of the Na,K-ATPase pump cycle is to choose conditions in which the pump can only undergo a part of the reaction cycle. For example, in the absence of extracellular K^+ but at high intra- and extracellular Na^+ concentrations (so called "Na^+/Na^+exchange conditions"), the enzyme is restricted to the Na^+ translocating limb of the pump cycle. Voltage-jumps under these conditions result in ATP-dependent, ouabain-sensitive transient (pre-steady state) currents which arise as a consequence of Na^+ charge-movements involving the Na^+-loaded, phosphorylated forms of the sodium pump. This has been interpreted in terms of a high-field extracellular "access channel" or "ion well" for Na^+ ions (Gadsby et al., 1993), since changes in membrane potential have been shown to be kinetically equivalent to changes in the concentration of sodium ions that bind within the ion well (Läuger, 1991): Elevation of the external Na^+ concentration shifts the Boltzmann-type Q/V curves to more positive potentials, since less hyperpolarizing potentials are required to saturate the extracellular binding sites at the bottom of the ion well (Sagar and Rakowski, 1994). High-speed voltage-jumps on internally dialysed squid axons revealed three distinct phases in the resulting presteady state charge movements, thus indicating that the de-occlusion and release of all three sodium ions is electrogenic and occurs in a strict sequential order (Holmgren et al., 2000). Considering these

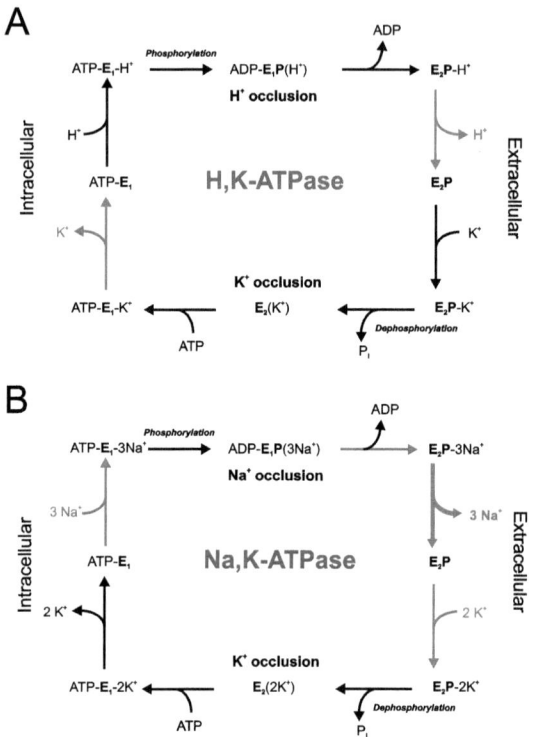

Figure 1.2.: **Electrogenic steps in the H,K-ATPase and Na,K-ATPase reaction cycles.** Steps in the reaction cycles of H,K-ATPase (A) and Na,K-ATPase (B) that were shown to be at least weakly electrogenic are indicated by grey arrows. For the Na,K-ATPase, the major electrogenic event was linked to the extracellular Na$^+$ release or reverse binding steps (represented by bold arrows in B)

three binding steps, it is the binding or release of the third Na$^+$ ion that represents the major electrogenic event. In contrast to the other two Na$^+$ ions that are presumably bound to residues which are also part of the binding sites for the two K$^+$ ions, the coordinating residues of the third Na$^+$ ion remain probably unligated in the K$^+$-translocating branch of the cycle. The lower voltage-dependence of the intracellular Na$^+$ and extracellular K$^+$ binding suggests rather low-field access channels for these weakly electrogenic steps. A low-field access channel is characterized by a wide opening (or vestibule) into which water and all kinds of ions may freely enter, resulting in a high conductance. Accordingly, only a small fraction of an externally applied voltage drops across the length of the channel, which is reflected by the lower dielectric coefficients of ∼0.1-0.25 compared to ∼0.7-0.8 for the high-field extracellular Na$^+$ access channel (Rakowski et al., 1997). However, it should be noted that apart from the ion access channel model described here, other theories may explain the electrogencity of ion binding and release steps as well. Being based on the idea of electrostatic barriers that move across the transmembrane electric field, these theories may explain the electrogenicity without any necessity for an access channel in molecular terms. Therefore, they represent attractive alternatives to the ion access channel theory, especially regarding the lack of structural evidence for ion access channels from the currently available Na,K-ATPase crystal structures (see next section).

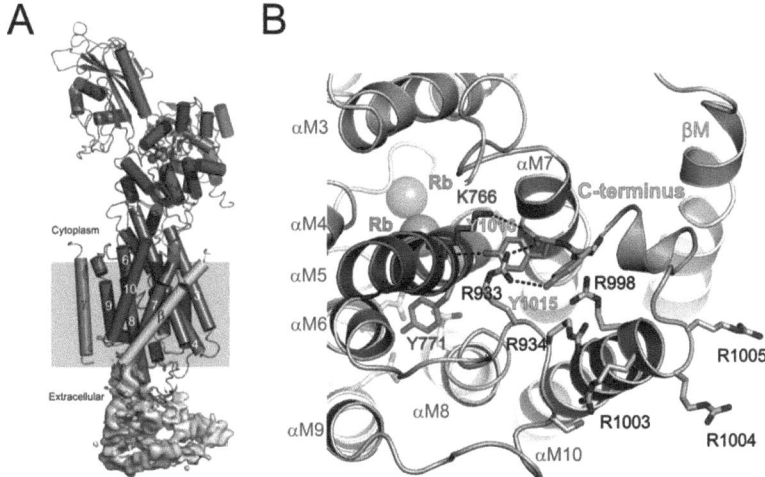

Figure 1.3.: **Crystal structure of the pig renal Na,K-ATPase.** A; overall view showing the oligomeric Na,K-ATPase complex of α_1-, β_1- and γ-subunit. B; close-up view from the cytoplasmic side, illustrating the C-terminal intrusion into the transmembrane region and possible interacting residues (e.g. K766 and R933) of the two C-terminal tyrosines Y1015 and Y1016. The two occluded Rb^+ ions in the cation binding pocket are shown as spheres. Both structural representations have been adapted from Morth et al. (2007).

The putative charge-translocating steps in the H,K- and Na,K-ATPase reaction cycles are summarized in Figure 1.2 A and B, respectively.

1.1.3. Structure of P-Type ATPases

The first high-resolution crystal structure of a P-type ATPase appeared in 2000 (Toyoshima et al., 2000), revealing the Ca^{2+}-bound E_1-state of the sarcoplasmatic reticulum ATPase (SERCA1a) from rabbit skeletal muscle at 2.6 Å. Shortly after, additional structures of SERCA1a in several different conformational states were resolved (Toyoshima and Nomura, 2002; Toyoshima and Mizutani, 2004; Olesen et al., 2004, 2007), thus indicating the structural changes that occur when the enzyme moves through the catalytic cycle. These P_{2A}-type ATPase structures are of a more general significance, since it is assumed that similar changes probably take place in related P-type ATPases. However, the structures of this monomeric ATPase were not able to provide any structural information about the interaction sites with accessory subunits of certain oligomeric P-type ATPases (see section 1.1.4).

Crystal Structure of the Na,K-ATPase at 3.5 Å resolution

Another major breakthrough came in 2007, when the X-ray structure of the pig renal Na,K-ATPase ($\alpha_1\beta_1\gamma$-complex, see Figure 1.3 A) with two bound Rb^+ ions at 3.5 Å resolution was published (Morth et al., 2007). Apart from revealing the association of the β-TM with α-TM7/α-TM10 (see Chapter 4 for details) and of the γ-TM with α-TM9, another interesting finding was the location of the carboxy terminus of the α-subunit. Compared to the SERCA1a, the C-terminus of the Na,K-ATPase carries an extension of eight residues which forms a short α-helix that is accommodated between β-TM, α-TM7 and α-TM10 (see Figure 1.3 B). The last two residues of the C-terminus are highly conserved tyrosines wich are present in all four Na,K-ATPase α-isoforms and the gastric and non-gastric H,K-ATPase. Notably, they seem to interact with positively charged residues (Lys-766 of α-TM5 and Arg-933 in the M8/M9 loop) according to the crystal structure. Since a C-terminal deleted variant (ΔKETYY) led to a 26-fold reduction of Na^+ affinity (a phenotype with strong similarity to the ones resulting

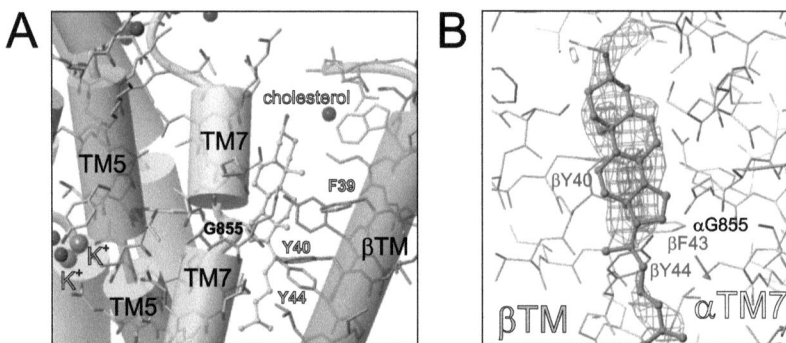

Figure 1.4.: **Crystal structure of the Na,K-ATPase from shark rectal gland at 2.4 Å resolution.** A, a close-up view of the α/β interface illustrating the partial unwinding of TM5 and TM7 of the Na,K-ATPase α-subunit (indicated by kinks in the cylindrical helices). Note the participation of cholesterol and aromatic β-TM residues in stabilizing the unwound part of TM7 (around Gly855). B, details of the cholesterol binding site, highlighting the aromatic residues of the β-TM involved in interactions to the steroid moiety of cholesterol. Mesh represent electron densities contoured at 2σ and 4σ, respectively. Both structural representations in A and B are adapted from Shinoda et al, 2009 (PDB code 2ZXE).

from mutations of the putative third Na^+ binding site (Li et al., 2005)), it was speculated that the C-terminus might be involved in binding of the third Na^+ ion. A cluster of positively charged arginines close to the C-terminus (Arg933, Arg934, Arg998, Arg1003, Arg1004, Arg1005; pig kidney α_1-isoform numbering) was suggested to function as a voltage sensor that controls positioning of the C-terminus.

Further biochemical studies on Na,K-ATPase α_1-mutants with C-terminal deletions of various length revealed that the C-terminus is not only critical for Na^+ binding from the cytoplasmic site, but also for Na^+ binding from the extracellular site, since high concentrations of Na^+ were unable to induce the $E_2P \rightarrow E_1P$ reverse reaction in these mutants. Interestingly, a similar phenotype was also observed when Arg933 (pig kidney α_1-isoform numbering) was mutated to alanine. Since this residue is in position to form one or two salt bridges or hydrogen bonds with the C-terminal carboxylate group, it may represent a functional and structural link between C-terminus and Na^+ binding site (Toustrup-Jensen et al., 2009).

The physiological relevance of these structural findings is underscored by the fact that a mutation of the corresponding arginine residue (to proline) in the Na,K-ATPase α_2-isoform or a C-terminal extension of this isoform by 28 amino acids was linked to familial hemiplegic migraine Type 2 (FHM2) (Riant et al., 2005). Notably, even a substantially shorter extension of the α_3-Na,K-ATPase C-terminus (by a single additional tyrosine residue) resulted in an approximately 50-fold reduction of the intracellular Na^+ affinity and was associated with sporadic rapid-onset dystonia-parkinsonism (RDP) (Blanco-Arias et al., 2009).

Finally, electrophysiological characterization of C-terminally deleted or mutated α_2-mutants demonstrated that the two terminal tyrosines are essential for a normal K^+-activated pump activity: In the absence of extracellular K^+ the ΔYY and $YY \rightarrow AA$ α_2-mutants mediated ouabain-sensitive, hyperpolarization-activated inward currents, which were predominantly carried by Na^+ and H^+, since they were dependent on extracellular Na^+ and increased at lower pH. For the wildtype Na,K-ATPase, similar inward currents were also observed at acidic pH, but only in the absence of both extracellular K^+ and Na^+ (Vasilyev et al., 2004). Therefore, the remarkable phenotype of these C-terminally altered mutants may be tentatively attributed to a disrupted intracellular gate that controls the strictly alternating access of cations to the ion binding pocket (Meier et al., 2009).

Crystal Structure of the Na,K-ATPase at 2.4 Å resolution

Although the aforementioned structure at 3.5 Å has already contributed significantly to a better understanding of the Na,K-ATPase at a molecular level, some important details including the mechanism of K^+ coordination by residues of the catalytic α-subunit were only revealed at a still higher resolution of 2.4 Å, which was achieved very recently by crystallization of the Na,K-ATPase $α_1$-isoform in the E_2-state from shark rectal glands (Shinoda et al., 2009).

Interestingly, it was shown that in contrast to Ca^{2+}-coordination by carboxylate side-chains of SERCA, mainly main-chain carbonyl oxygens contribute to the K^+ binding pocket of the Na,K-ATPase, thus possibly representing the key element for defining K^+ selectivity of the ion pump. Notably, the involvement of main-chain carbonyls in cation coordination requires a partial unwinding of the helical conformations of TM5 and TM7, which in turn implies additional structural particularities. Whereas the unwinding of TM5 is probably facilitated by the presence of a helix-breaking proline (Pro785, shark rectal gland $α_1$-isoform numbering, corresponding to a glycine residue in the Ca^{2+} ATPase), the unwinding of TM7 most likely requires assistance of the β-subunit, since hydrogen-bonding between the backbone carbonyl of Gly855 in TM7 and Tyr44 in the β-TM stabilizes the unwound structure of TM7 (illustrated in Figure 1.4 on the facing page A). Additional stabilization might be achieved by a cholesterol molecule that appears to shield the unwound part of TM7 from exposure to bulk lipids of the membrane (see Figure 1.4 on the preceding page B). Apart from these interesting findings, the structure by Toyoshima and colleagues highlighted key residues that are responsible for the interactions between β-ectodomain and α-subunit and also some contact sites to the accessory protein FXYD (see upcoming section 1.1.4). Another noticeable observation was that the majority of the residues involved in these protein interactions are aromatic amino acids.

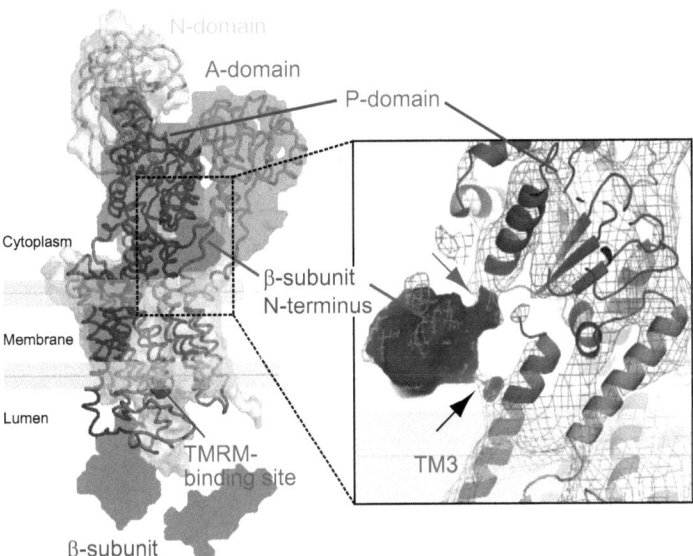

Figure 1.5.: **Structural representation of the pig gastric H,K-ATPase based on the recently published cryo-EM structure.** The cryo-EM structure (EM Data Bank code 5104) is represented by surface or mesh, contoured at 1 σ. A homology model based on the Na,K-ATPase structure is also shown (cartoon, PDB code 3IXZ). Inset: a close-up view (from the right side of the molecule) showing the putative interaction sites of the β-subunit N-terminus with the P-domain (upper arrow) and TM3 (lower arrow) of the α-subunit, respectively. *(This figure was prepared by K. Abe, University of Kyoto, Japan.)*

Cryo-EM Structure of the gastric H,K-ATPase

Despite the aforementioned recent progress in the available structural information for related P-type ATPases, no high-resolution X-ray structure of the gastric H,K-ATPase has been published so far. However, electron microscopy of two-dimensional crystals recently revealed the three-dimensional structure of the pig gastric proton pump (complexed with aluminium fluoride in the pseudo-E_2P conformation) at 6.5 Å resolution (Abe et al., 2009).

An interesting finding of this study was the close proximity between the short cytosolic N-terminal tail of the β-subunit and the P-domain of the α-subunit (see red arrow in Figure 1.5, inset). The authors suggested that this putative inter-subunit contact may tether the P-domain in an E_2P-specific position by a 'ratchet'-like mechanism, thereby preventing the $E_2P \rightarrow E_1P$ reverse reaction. *In vivo*, this could facilitate proton release against the million-fold proton gradient of the stomach lumen. Notably, as indicated by the black arrow in Figure 1.5, the EM structure proposed another possible interaction between β-N-terminus and the α-subunit near the cytoplasmic part of TM3, but it is less clear whether this contact influences the E_1P/E_2P conformational distribution of the α-subunit as well (see chapter 5 for further details).

1.1.4. Oligomeric P-Type ATPases

Essential ancillary subunits (β-subunits)

Only a few members of the P-type ATPase family require accessory subunits in addition to the main ATP-hydrolyzing subunit for transport activity. For a long time, it was assumed that the need for additional subunits was a unique feature of all K^+-countertransporting P-type ATPases, since the Kdp ATPase subunits KdpA, KdpC, KdpF and the Na,K- and H,K-ATPase β-subunits were the only known essential accessory subunits in the whole P-type ATPase superfamily. The transport of potassium ions might require particular sequence information which, however, might complicate the proper membrane integration of the respective transmembrane parts. The accessory subunits may have evolved as helper proteins that facilitate the more demanding task of folding and maturation. For example, KdpC was shown to assist in the assembly of the Kdp ABCF complex (Gassel and Altendorf, 2001). Therefore, it was hypothesized that the β-subunits of Na,K- and H,K-ATPases could be remnants of the bacterial KdpC subunit that have been eliminated in other (non K^+-transporting) ATPases (Geering, 2001). Although results from baculovirus-infected insect cells have shown that at least the Na,K-ATPase α-subunit alone is able to perform ATP hydrolysis (Blanco et al., 1994a), the β-subunit is essentially required for ion transport of Na,K- and H,K-ATPase enzymes. Furthermore, in *Xenopus* oocytes and mammalian cells, the proper folding, maturation and plasma membrane targeting of the Na,K- and H,K-ATPase α-subunits critically depend on the presence of the β-subunit. (For more detailed information on the functional significance of the Na,K- and H,K-ATPase β-subunits, see reviews by Geering (2001) and Chow and Forte (1995) or the introductory parts of chapters 3-5).

Yet, more recently, a family of putative β-subunits of Type-IV ATPases was identified in yeast (Saito et al., 2004) with homologues in plants (Poulsen et al., 2008a) and humans (Katoh and Katoh, 2004). Interestingly, these cdc50 proteins consist of two transmembrane helices and structurally resemble a fusion protein between β- and γ-subunits of the Na,K-ATPase (see below) in terms of polypeptide length, N-glycosylation and membrane topology. In analogy to the β-subunits of Na,K- and H,K-ATPase, the cdc50-like proteins are involved in trafficking of the respective catalytic Type IV-ATPase and probably contribute either directly to the formation of a phospholipid binding site or indirectly by inducing conformational changes in the 10-helix bundle of the P_{IV}-ATPase α-subunit, thereby creating a high affinity phospholipid binding site (Lenoir et al., 2007).

Notably, very recent results showed that the affinity of cdc50 for its binding partner (the yeast P_{IV}-type ATPase Drs2p) fluctuates during the transport cycle. Remarkably, the strongest interaction was observed in the E_2P-state, in which loading with the phospholipid substrate takes place (Lenoir et al., 2009). The authors therefore concluded that the E_2P stabilization mediated by the cdc50 interaction may increase the loading efficiency of the flippase which is presumably limited by the slow two-dimensional diffusion rate of the phospholipid substrates in the plasma membrane (Puts and Holthuis, 2009). Similar to the proposed ratched-like function of the gastric H,K-ATPase β-subunit which is assumed to promote H^+ release from the E_2P-state against the enormous pH gradient *in vivo* (see section 1.1.3 and chapter 5 on page 67), the prolonged dwell time in E_2P would enhance the phospholipid binding efficiency of the flippase.

Non-essential, regulatory subunits (γ-subunits)
Recently, a class of small single-span transmembrane proteins associated to the Na,K-ATPase was identified, the so-called FXYD family, named after a common signature sequence FXYD prior to the transmembrane segment (Sweadner and Rael, 2000). Although not being required for basic transport function itself, all seven members of this family (FXYD1-7) were shown to modulate the transport properties of the pump, e.g. by causing small alterations in turnover number or ion affinities. Since each member is only expressed in certain specialized tissues (see Table 1.2), this proteins have probably a regulatory role in fine-tuning the Na,K-ATPase transport activity to the specific needs in these tissues. For instance, the association of the kidney-specific γ-subunit (FXYD2) with α- and β-subunits of the Na,K-ATPase was shown to influence the enzyme's apparent affinities for ATP, Na^+ and K^+ (probably as a consequence of an shift in the conformational E_1/E_2 equilibrium towards E_1) and the voltage-dependence of K^+ activation in both mammalian cells (Therien et al., 1997, 1999, 2001) and *Xenopus* oocytes (Beguin et al., 1997). Since the apparent affinity for ATP was about two-fold higher in γ-associated ATPase complexes isolated from HEK293 cells, it was speculated that this kidney-specific modulation of the Na,K-ATPase's transport properties may be crucial under energy-compromised conditions in some putative anoxic parts of the renal outer medulla (Therien et al., 1999). The physiological significance of the γ-subunit in the kidney is underscored regarding the hereditary hypomagnesemia disorder HOMG2 that was linked to an autosomal dominant negative mutation (G41R) in the γ-TM region (Meij et al., 1999). The phenotype of these patients is possibly explained by a routing defect of the γ-subunit that presumably affects the whole αβγ-Na,K-ATPase complex (Meij et al., 2000), thus reducing Na,K-ATPase mediated K^+ import into the cell. As a result, the intracellular K^+ concentration drops substantially which in turn affects Mg^{2+} reabsorption via a K^+-dependent Mg^{2+} entry channel in the apical membrane of the renal tubule cells.

Notably, another tissue-specific modulation of Na,K-ATPase activity is mediated by phospholemman (PLM or FXYD1) in heart and muscle cells. Binding of PLM to the Na,K-ATPase results in a 2-fold reduction of the intracellular Na^+ affinity (Crambert et al., 2002), but the effect is lost when PLM is phosphorylated by protein kinase A (e.g after β-adrenergic stimulation by epinephrine, as described by Bibert et al., 2008). In cardiac myocytes, this modulation of the Na^+ affinity could be relevant for an efficient extrusion of intracellular Na^+, thus increasing Ca^{2+} efflux via the Na^+/Ca^{2+} exchanger. This is probably of utmost importance for avoiding Ca^{2+} overload-induced arrhytmias and for limiting positive inotropy during sympathetic stimulation (Despa et al., 2008).

Interestingly, the related single-span membrane protein phospholamban (PLB) that however does not belong to the FXYD family, was shown to modulate the transport activity of the sarcoplasmatic reticulum Ca^{2+}-ATPase (SERCA2a) in heart cells by a similar mechanism (Simmerman et al., 1986): Binding of PLB to SERCA2a reduces the Ca^{2+} affinity (Cantilina et al., 1993), resulting in an inhibitory effect on the protein's Ca^{2+} transport activity (James et al., 1989). Upon β-adrenergic stimulated phosphorylation of PLB by protein kinase A, the inhibitory interaction is disrupted and SERCA-mediated Ca^{2+} transport is about 3-fold enhanced. Notably, a single missense mutation in a *PLB* allele (R9C) was shown to be responsible for a dominant inherited form of dilated cardiomyopathy and heart failure (Schmitt et al., 2003). The mutation results in chronic SERCA2a inhibition, thereby causing Ca^{2+} overload-induced arrhytmias and heart failure. Significantly, the mutation-carrying PLB is not directly inhibiting SERCA2a (e.g. by being resistant to PKA phosphorylation-induced release from the pump). Instead, due to an increased affinity for PKA, the mutated PLBs are unable to dissociate from PKA, thus rendering the kinase unable to phosphorylate wild type PLB molecules, which results in constitutively active wild-type PLB. This explains the dominant phenotype of this heterozygous genetic defect. Since intracellularly elevated Ca^{2+} levels are common in most cardiomyopathies, PLB antisense RNA or gene transfer of a constitutively phosphorylated PLB mutant (S16E) may be promising new therapeutic approaches in the treatment of these diseases (MacLennan and Kranias, 2003).

isoform	tissue distribution
α_1	all tissues
α_2	adipocytes, muscle, heart, brain (glia cells)
α_3	neurons, heart
α_4	testis
β_1	all tissues
β_2 (AMOG)	nervous tissues (glia cells), pinneal gland, sceletal muscle
β_3	testis, retina, liver, lung
γ (FXYD2)	kidney
FXYD1 (PLM)	heart, sceletal muscle
FXYD3 (Mat-8)	uterus, colon, stomach (mucus cells), cancer cells
FXYD4 (CHIF)	kidney, colon
FXYD5 (Ric)	kidney, intestine, lung
FXYD6 (phosphohippolin)	inner ear, brain, lung, testis, colon
FXYD7	brain

Table 1.2.: **Tissue-specific expression of different Na,K-ATPase α-, β- and γ-isoforms.**

Multimer Formation of dimeric Na,K- and H,K-ATPase

Interestingly, both Na,K-ATPase and gastric H,K-ATPase α/β heterodimers probably form higher order oligomers with important functional consequences for enzymatic activity under physiological conditions:

Several biochemical studies indicate that the functional form of the H,K-ATPase is actually an α_2/β_2 oligomer (Morii et al., 1996; Shin and Sachs, 1996; Abe et al., 2002, 2003; Shin and Sachs, 2004; Abe et al., 2005). The native form of the proton pump probably functions as an "out of phase" oligomeric heterodimer: when one α/β heterodimer is in the E_1-form, the other one is obligated to be in the E_2-form, thereby allowing the large conformational changes in the cytoplasmic domain of the α-subunit that occur during the $E_1 \rightarrow E_2$ transition. The close association of the two α-subunits in the tetrameric α_2/β_2 form would most likely prevent a simultaneous E_1 conformation of both catalytic subunits, assuming identical movements of N- and A-domain to those in the SR Ca^{2+}-ATPase (Toyoshima et al., 2000). The formation of such an "out of phase" oligomer has been concluded from measurements of the stoichiometry of ATP binding, acid-stable phosphorylation and binding of inhibitors (Eguchi et al., 1993; Abe et al., 2002; Shin and Sachs, 2004; Shin et al., 2005). The α_2/β_2 heterodimer would also explain why two ATP binding sites with different affinities are characteristic for the H,K-ATPase (Wallmark et al., 1980): A high affinity binding site ($K_{0.5}$ in the submicromolar range) and a low affinity site with a $K_{0.5}$ in the submillimolar range (Helmich-de Jong et al., 1986). Since these two sites are interconvertible, they presumably correspond to ATP binding sites in the E_1- and E_2-states, respectively. ATP binding to the low affinity site in the E_2-state was not only shown to accelerate the transition to E_1 (analogous to the effect of low affinity ATP binding to the Na,K-ATPase (Robinson, 1967; Hegyvary and Post, 1971)), but also prevents the formation of an inhibitory $E_1 K$-state at physiologically relevant high intracellular K^+ concentrations ($\sim$150 mM). Of note, cytosolic inhibition by K^+ delays phosphoenzyme formation (and H^+ secretion) at low ATP, but not at high ATP levels, where both sites are occupied (Lorentzon et al., 1988). This illustrates the possible benefit of a oligo-heterodimeric H,K-ATPase complex under *in vivo* conditions.

Interestingly, the formation of an oligomeric heterodimer was also proposed for the Na,K-ATPase by numerous studies, applying a diversity of techniques (Stein et al., 1973; Repke and Schön, 1973; YuA et al., 1985; Pachence et al., 1987; Ting-Beall et al., 1990; Tsuda et al., 1998; Taniguchi et al., 1999, 2001; Kaya et al., 2003; Laughery et al., 2004; Clarke and Kane, 2007; Pilotelle-Bunner et al., 2008). Although the arrangement of Na,K-ATPase molecules in high-resolution crystals rather suggests a monomeric heterodimer (Morth et al., 2007) and several studies on purified Na,K-ATPase enzymes also argue for an α/β protomer as the minimum functional enzyme unit (Brotherus et al., 1983; Vilsen et al., 1987; Hayashi et al., 1989), diprotomer formation was observed in two-dimensional crystals by electron microscopy (Yokoyama et al., 1999). Furthermore, the existence of oligomeric heterodimers under more physiological conditions (i.e. in biological membranes) seems reasonable, since it would actually explain the stoichiometry of ligand binding (Buxbaum and Schoner, 1991; Yokoyama et al., 1999; Taniguchi et al., 2001) and the phosphorylation kinetics under certain conditions (Fröhlich et al., 1997). Moreover, several studies have revealed regions in both α- and β-subunit involved in α/α- (Blanco et al., 1994b; Koster et al., 1995; Zolotarjova et al., 1995; Donnet et al., 2001) or β/β interactions (Ivanov et al., 2002; Barwe et al., 2007), respectively. Recently, Clarke and Kane suggested a "two gear model" for the sodium pump: a "low gear" with one ATP bound per diprotomer and a "high gear" with two ATP molecules bound. Whereas the low gear mode operates at low ATP concentrations, a switch to the high gear (at higher ATP concentrations) would result in enhanced turnover number. This may be relevant in excitable cells, such as nerve and muscle cells, where a faster ion pumping rate by the Na,K-ATPase would enable the cell to return to its resting potential in a shorter period of time after excitation. This in turn would save ATP, since during this shortened time period less passive leakage of ions occurs that must be counteracted by the pump (Clarke and Kane, 2007).

Notably, electron microscopic studies hint at the formation of even higher oligomeric Na,K-ATPase complexes (i.e. $(\alpha/\beta)_4$) (Maunsbach et al., 1991; Yokoyama et al., 1999), which are actually in line with results from early cross-linking studies that suggested a minimum of four catalytic subunits in Na,K-ATPase oligomers (Askari and Huang, 1980). Moreover, it was shown that solubilized Na,K-ATPase occurs in a mixture of protomers (150 kDa), diprotomers (300 kDa) and tetraprotomers (Hayashi et al., 1997; Yokoyama et al., 1999; Kobayashi et al., 2007; Mimura et al., 2008). Likewise, 2 D crystals of membrane bound enzyme indicate that also the gastric H,K-ATPase is associated in tetraprotomeric complexes (Hebert et al., 1992), which is further supported by results from a study using high performance gel chromatography and total internal reflection microscopy (Abe et al., 2003). A modified Albers-Post scheme for the Na,K-ATPase reaction cycle that explicitly accounts for this putative tetraprotomeric structure was presented by Taniguchi (Taniguchi et al., 1999, 2001).

1.2. Physiological Roles of Na,K- and H,K-ATPases

1.2.1. Physiological Functions of the Na,K-ATPase

Regulation of Cell Volume

The Na,K-ATPase establishes and maintains the high internal potassium and low internal sodium concentrations characteristic of most animal cells. Its ubiquitous occurence is probably a direct consequence of the fundamental importance of these ion gradients to basic cellular functions such as regulation of osmolarity and cell volume. Persistent inhibition of the sodium pump is fatal for almost all kinds of cells since they contain a substantial amount of anionic colloids (mostly proteins and organic phosphates) that cannot cross the membrane and therefore attract small cations from the interstitial fluid (according to the Gibbs-Donnan equilibrium). If this is not actively counteracted by the Na,K-ATPase which pumps Na^+ ions out of the cell, the osmolarity of the cell interior exceeds the extracellular one and water is continously drawn into the cell. The resulting osmotic swelling of the cell would eventually result in rupture of the plasma membrane and necrosis of the sodium pump-deficient cell.

Resting Potential and Secondary Active Transport

Apart from this essential basic function in ionic balance of all cells, the sodium pump is also highly important for some cellular functions of more specialized cell types. Nutrient uptake is often mediated by secondary active transporter that utilize the Na^+ or K^+ gradients which are maintained by the sodium pump. The most famous example is the Na^+/glucose cotransporter SGLT1 located in the mucosa of the small intestine (Wright et al., 1992). Another important member of this family, SGLT2 is responsible for renal glucose reabsorption in the proximal tubule of the nephron (Wright, 2001). Furthermore, many amino acids as well as metabolic intermediates (such as malate, succinate etc.) are transported via Na^+-driven symporters of the major facilitator superfamily (MFS). The fact that in excitable or kidney cells up to one third of the ATP consumption can be ascribed to Na,K-ATPase activity, highlights the significance of the sodium pump in these particular cell types. The high level activity in excitable cells is not only required to ensure a rapid return to the resting potential and thus enable rapidly recurring excitation of neuron or muscle cells, but also to limit the duration of electric pulses: The removal /reuptake of many neurotransmitter from the synaptic cleft is carried out by members of the neurotransmitter sodium symporter family (NSS) that rely on ionic gradients established by the sodium pump.

Na,K-ATPase isoforms

The existence of various tissue-specific α-, β- and γ- subunit isoforms that are predominantly expressed in cells with extraordinarily high Na,K-ATPase activity (see Table 1.2) strongly suggests that an additional fine-tuning of the transport properties is necessary to fullfill the particular needs of these cell types (see also previous section on FXYD isoforms). Although the sequence differences in the various α-isoforms are only minor, the resulting subtle changes in the affinities for cations and ATP may become relevant under certain physiological conditions (Blanco and Mercer, 1998). For instance, the $α_2$-isoform has a slightly higher intracellular Na^+ affinity (probably as a consequence of a shift toward E_1 (Segall et al., 2001)) which is advantageous under low intracellular Na^+ concentrations that are characteristic for non-excitable glial cells. High level activity of the sodium pump in these cells is crucial for efficient removal of extracellular potassium to prevent further depolarization of surrounding neuronal cells. The fact that several mutations in the *ATP1A2* gene encoding the human $α_2$-isoform were linked to an autosomal dominant inherited form of migraine (familial hemiplegic migraine type-2, FHM2) highlights the importance of this tissue-specific isoform (De Fusco et al., 2003; Moskowitz et al., 2004; Riant et al., 2005; Tavraz et al., 2008, 2009).

Another example is the $α_3$-subunit which is uniquely expressed in excitable neuronal tissues and has a relatively low affinity for cations. It may help to restore the resting membrane potential after repeated firing of action potentials when the Na^+ and K^+ gradients are dissipated. Under these conditions, the $α_1$- and $α_2$- isoforms (that are also present in neurons) are already working under saturation, but the $α_3$-isoform is well below saturation and may therefore function as a "spare pump". Furtermore, its increased affinity for ATP enables a high turnover number even at the low ATP concentrations that are typical near the cell membrane after intense neuronal activity. Notably, several missense mutations in the *ATP1A3* gene were shown to be associated with Rapid-Onset Dystonia Parkinsonism (de Carvalho Aguiar et al., 2004; Blanco-Arias et al., 2009).

Finally, slight differences of the various isoforms regarding their apparent affinity for cardiac glycosides (or regarding the K^+ antagonism of these inhibitors, see Crambert et al. (2000)) as well as their different subcellullar localization are assumed to be responsible for the differential effects of these drugs on $α_2$- and $α_3$-isoforms in heart cells (see the following section).

1.2.1.1. Na,K-ATPase Inhibitors

Already centuries before the discovery of the sodium pump by Jens C. Skou in 1957 (Skou, 1957), its inhibitors were well known for their poisonous effects. The inhibitory compounds belong to a class of cardiac glycosides which are composed of a sugar (glycoside), a steroid and a lacton (a cyclic ester) moiety (see Figure 1.6). About 200 members of this class of inhibitors are known, most of them are found in plants but some are also isolated from animals (toads and snakes). Ouabain, for example, is a well-known East-african arrow poison that was usually obtained from the seeds of different plants belonging to the *Acokanthera* or *Strophanthus* species. Today, the most important source of these drugs are various *Digitalis* species (e.g. foxglove). For more than two centuries, digitalis glycosides have been used in the treatment of congestive heart failure and cardiac arrhythmia, but only in the last decades, the molecular details of their positive inotropic effects have begun to be elucidated.

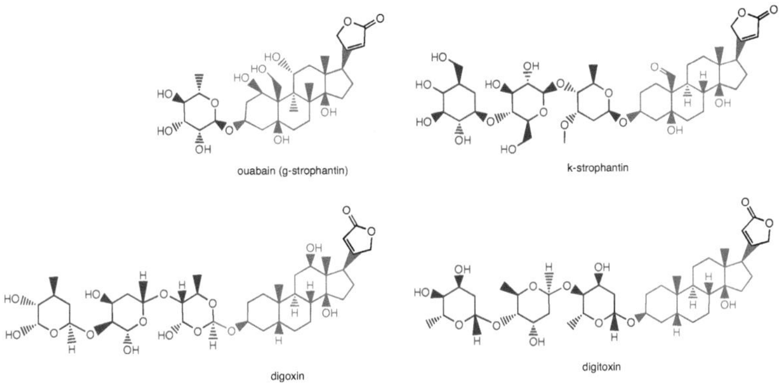

Figure 1.6.: **Chemical structures of cardiac glycosides.** Two cardiac glycosides from *Strophantus* (upper row) and two from *Digitalis* (lower row) are shown.

Ouabain and all other cardiac glycosides specifically bind with high affinity to the extracellular side of the E_2P conformer (Yoda and Yoda, 1982), thereby locking the enzyme in this state and preventing the $E_2 \rightarrow E_1$ transition. The precise location of the high-affinity ouabain binding site is not known yet (see below), but numerous studies have narrowed it down to a couple of amino acids in extracellularly located parts of TM4, TM5 and TM6 (Palasis et al., 1996; Koenderink et al., 2000; Qiu et al., 2005, 2006). In addition, two amino acids at the edge of the TM1/TM2 extracellular loop were shown to be responsible for species-specific variations in ouabain affinity: the α_1-isoform of rodents has a 1000-fold lower affinity than their α_2- and α_3-isoforms or the α_1-subunits of all other species, which was attributed to charged residues in these positions (Arg111 and Asp122 instead of Gln111 and Asn122)(Price and Lingrel, 1988). Despite these differences in ouabain sensitivity observed between rodent α-isoforms, all human isoforms appear to have similar affinities for ouabain (Wang et al., 2001). This, however, challenges many models that explain the effects of cardiotonic steroids on the basis of a selective inhibition of certain subsets of pumps in the heart, thus relying on isoform-specific differences in ouabain sensitivity. A famous model by Blaustein and coworkers (Juhaszova and Blaustein, 1997) suggests a selective inhibition of α_2 pumps that are located in the so-called "PLasmERosome", consisting of plasma membrane microdomains and adjacent junctional ER that enclose a tiny volume of cytosol. This spacial arrangement brings the α_2-ATPases not only in close proximity to the Na^+/Ca^{2+}transporter (NCX) in the plasma membrane, but also to SERCA molecules of the sarcoplasmic membrane. This would enable a local elevation of cytoplasmic $[Na^+]$ and $[Ca^{2+}]$ and enhanced positive inotropic Ca^{2+} signalling without concomitant changes in the bulk cytosolic ion concentrations that were actually shown to be absent in α_2-deficient astrocytes (Golovina et al., 2003).

Crystal structure of the Na,K-ATPase with bound ouabain

Very recently, a high resolution crystal structure of the Na,K-ATPase from shark rectal gland with bound K^+ and ouabain (i.e. $E_2 \cdot 2K \cdot P_i$) was published, revealing interesting details about the interactions with the compound (Ogawa et al., 2009). Contrary to previous models that placed ouabain on the extracellular surface, the structure at 2.8 Å resolution showed that ouabain is wedged deeply into the transmembrane cleft, resulting in a partial unwinding of the TM4 helix which could explain why ouabain binding is relatively slow. Apart from Π-stacking between three aromatic phenylalanine residues (in TM4 and TM5) and the steroid core of ouabain, charged residues that form hydrogen-bonds to the sugar moiety of the compound contribute additionally to the strength of the interaction (see Figure 1.7). However, due to the long-known K^+-antagonism of ouabain binding, the state captured in this X-ray structure represents a low affinity ouabain-bound state. Therefore, in the high-affinity ouabain bound state, strong additional contacts must be present which are not revealed by this structure. A homology model based on the E_2-BeF_3^- crystal structure of Ca^{2+}-ATPase (corresponding to E_2P without K^+) was used to predict the interactions that contribute to the high-affinity binding. The homology modeling performed by Ogawa and coworkers suggests an additional hydrophobic binding cavity that is mainly formed by residues in TM1 and TM2. This complementary surface is assumed to be only available when no K^+ ions are present in the cation binding pocket, thereby significantly enhancing the binding affinity of ouabain towards the E_2P state in absence of the ionic antagonist.

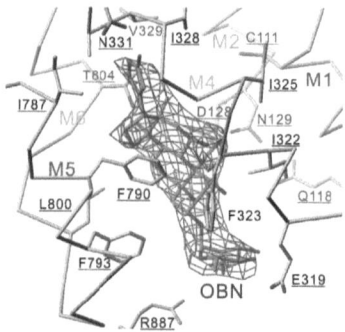

Figure 1.7.: **A close-up view of the low-affinity ouabain binding site of Na,K-ATPase from shark rectal glands.** The illustration was taken from Ogawa et al. and highlights the three phenylalanines involved in binding of the steroid core (Phe323 on TM4, Phe790 and Phe793 on TM5) and the two charged residues (Arg887 and Glu319) that stabilize the sugar moiety of ouabain (OBN) by hydrogen-bonding.

1.2.2. Physiological Functions of H,K-ATPases

1.2.2.1. The nongastric H,K-ATPase

It should be mentioned that apart from the gastric H,K-ATPase, which is the focus of the following sections, a so-called "nongastric" H,K-ATPase exists that is expressed in several tissues of the gut, including bladder (Burnay et al., 2001), distal colon (Cougnon et al., 1996; Takahashi et al., 2002) and the collecting duct (Buffin-Meyer et al., 1997). Although initially described as H,K-ATPases on the basis of Rb^+ uptake, extracellular acidification and intracellular alkalization experiments that suggested K^+/H^+ countertransport (Jaisser et al., 1993), it was later shown that these "H,K-ATPases" also transport Na^+ in exchange for K^+ (Cougnon et al., 1998; Grishin and Caplan, 1998; Codina et al., 1999). Moreover, they behave rather like Na,K-ATPases regarding their pharmacological properties: Most of them were shown to be sensitive only towards ouabain (Cougnon et al., 1996; Rajendran et al., 2000), but some have been reported to be also sensitive towards high concentrations of SCH 28080 (Jaisser et al., 1993; Buffin-Meyer et al., 1997). Interestingly, ouabain-sensitive transport of ammonium ions was demonstrated for the nongastric H,K-ATPase in *Xenopus* oocytes (Cougnon et al., 1999) which may be of physiological relevance in the kidney medulla.

1.2.2.2. Localization and Regulation of gastric H,K-ATPase

Almost all vertebrates have acid-producing cells in their gastric mucosa. Whereas in submammalian species, secretion of acid and pepsinogen are combined in one cell type, referred to as oxyntic cell (οξύσ = Greek for "acid"), these two functions are separated into two different cell types for the mammalian species: acid secreting parietal cells that express the gastric H,K-ATPase and chief cells (zymogen or peptic cells) that are responsible for the synthesis, storage and secretion of pepsinogen. Moreover, two additional cell types with secreting activity are found in the gastric mucosa: mucus-secreting goblet cells and histamine producing endocrine cells (e.g. ECL or mast cells). The different cell types of the gastric mucosa are illustrated in Figure 1.8).

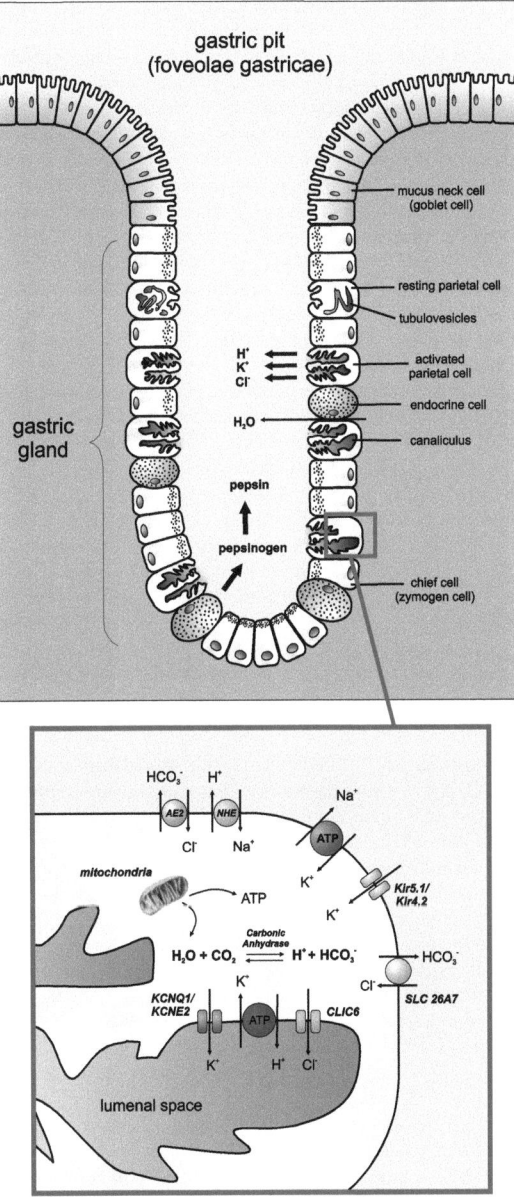

Figure 1.8.: **Schematic illustration of a gland in the gastric mucosa layer of the stomach.** Inset: activated parietal cell with the most important transporter and channel proteins involved in the formation of gastric fluid.

Upon stimulation of activating receptors on the basolateral surface of the parietal cell (typically by binding of acetylcholine, gastrin or histamine which in turn activates second messengers, such as cAMP and intracellular Ca^{2+}), the apical surface undergoes large morphological alterations which probably result from the fusion of H,K-ATPase containing cytoplasmic tubulovesicles with the rudimentary microvilli of the resting parietal cell. This leads to the formation of the elongated microvilli, the so-called secretory canaliculi that are characteristic for activated parietal cells (Forte et al., 1983).

Several lines of evidence indicate that this process of recruiting functional pumps to the cell surface does not involve the standard/regular exocytotic machinery of vesicle fusion. First, the proton pump was shown to reside in intracellular microtubules, not in vesicles (Ogata and Yamasaki, 2000). Moreover, the weak immunochemical staining of several proteins of the SNARE family (Sachs et al., 2007) suggest that these proteins are scarcely expressed in parietal cells which is further supported by the low amounts of mRNA for SNARE proteins (Lambrecht et al., 2005, 2006; Sachs et al., 2007). Importantly, the H,K-ATPase is restricted to the apical membrane of activated parietal cells. It was shown that the N-linked glycans attached to the highly conserved glycosylation sites on the H,K-ATPase β-subunit are responsible for proper targeting of mature H,K-ATPase complexes (Vagin et al., 2003, 2004, 2005b) (see also chapter 3).

The morphological changes that occur upon activation of parietal cells are however strictly reversible - cessation of acid secretion is associated with reinternalisation of the proton pump into intracellular storage compartments. Interestingly, the cytoplasmic tail of the H,K-ATPase β-subunit includes a sequence motif homologous to tyrosine-based endocytosis signals (FRHY). This motif is required for the internalisation of the proton pump and termination of acid secretion, since transgenic mice expressing a mutant β-subunit (Y20A) constitutively express the H,K-ATPase at the parietal cell surface and their parietal cells lack tubulovesicular storage compartments (Courtois-Coutry et al., 1997). However, despite the inability of the proton pump to be endocytosed, these HKβY20A-mice did not exhibit an increased basal acid secretion and the H^+ secreting activity in response to histamine was very similar to wildtype mice. This indicates that control of H,K-ATPase activity can occur independently of intracellular trafficking (Nguyen et al., 2004). This in turn strongly suggests that once present in the secretory canaliculi of the activated parietal cell, high-level H^+ secreting activity of the H,K-ATPase crucially depends on additional factors:

Luminal Potassium

Even before an ouabain-resistant, K^+-stimulated ATPase was identified in microsomes from gastric mucosa (Forte et al., 1967), it was a well-known fact that the presence of lumenal K^+ is an important prerequisite for H^+ secretion of the parietal cell (Harris et al., 1958). Although not much later it was demonstrated that stimulation of parietal cells not only triggers H^+ secretion, but also K^+ and Cl^- conductances in the apical membrane (Wolosin and Forte, 1984; Reenstra and Forte, 1990), the precise molecular nature of these transport pathways has only recently been revealed. CLIC6 (parchorin) appears to be the necessary accompanying Cl^- channel (Mizukawa et al., 2002) resulting in the observed net secretion of HCl. In 2001, two independent laboratories not only identified KCNQ1 as the missing lumenal K^+ channel in the parietal cell, but also showed its association with the β-subunit KCNE2 (Dedek and Waldegger, 2001; Grahammer et al., 2001). The fact that the KCNQ1-specific inhibitor chromanol 293B almost completely blocked acid secretion *in vivo* and *in vitro* (Grahammer et al., 2001) and that KCNQ1 knockout mice displayed hypochlorhydria (impaired gastric secretion), high plasma gastrin levels and hypertrophy of gastric mucosa (Lee et al., 2000; Vallon et al., 2005) highlights the crucial importance of these K^+ channels for acid secretion. Furthermore, a similar phenotype was observed for KCNE2 knockout mice (exhibiting severely impaired gastric acid secretion, abnormal parietal cell morphology, hypergastrinemia and glandular hyperplasia, as described by Roepke et al., 2006). KCNQ1/KCNE2 complexes were shown to be activated by acidic pH, PIP_2, cAMP and purinergic receptor stimulation (Heitzmann et al., 2004). This is highly interesting with regard to the fact that the latter three factors are long known activators of gastric acid secretion, but (apart from the Ca^{2+}-induced relocalization) H,K-ATPase activity itself is not regulated via these signalling pathways, eg. via phosphorylation by effector kinases. Since the typically low lumenal K^+ concentrations limit the turnover of the pump drastically, this indirect regulation of the H,K-ATPase is highly effective, thus explaining the relatively "mild" phenotype of transgenic βY20A-mice with constitutively apical targeted proton pumps.

Additional K$^+$ channels belonging to a class of inward rectifying potassium channels are found in the basolateral membrane of the parietal cell (Lambrecht et al., 2005). These Kir5.1 (KCNJ16)/Kir4.2 (KCN15) channels are assumed to replenish the intracellular K$^+$ concentration, which is reduced as a result of sustained K$^+$ efflux via activated KCNQ1/KCNE2 channels during ongoing gastric activity.

Cytosolic ATP

H$^+$ secretion of the gastric H,K-ATPase is an energetically highly demanding process. At maximum proton gradients with a ΔpH of approximately 6, the amount of free energy released by the hydrolysis of one ATP molecule ($\sim$50 kJ/mol under physiological conditions) is not sufficient to pump more than one proton per reaction cycle (Chow and Forte, 1995). It is therefore assumed that the proton pump adjusts its stoichiometry to different ΔpH.

To provide the enzyme with the required large amounts of ATP, parietal cells are extremely rich in mitochondria. About one third of the cytoplasmic volume is occupied by mitochondria (Helander et al., 1986), which is more than in almost all other types of cells (only the myocytes of the left cardiac ventricle have about the same proportion of mitochondria).

Cytosolic H$^+$

The H,K-ATPase is supplied with cytoplasmic protons by carbonic anhydrase, an enzyme which is present in the parietal cell at concentrations that are up to six time higher than in blood cells (Davenport, 1939). According to Sugai and Ito (1980), it is localized to both basolateral and apical membranes, in particular to the cores of the microvilli that project into the secretory canaliculus. This enzyme catalyzes the interconversion of water and carbon dioxide to a proton and a bicarbonate ion :

$$CO_2 + H_2O \rightleftharpoons H_2CO_3 \rightleftharpoons H^+ + HCO_3^-$$

Of course, like any other enzyme, it has no influence on the poise of the equilibrium. Therefore, a high metabolic activity of the parietal cell is crucial for the availability of protons, yielding high amounts of both educts (H$_2$O is produced via the reduction of oxygen in the mitochondrial respiratory chain and CO$_2$ is provided by the cytosolic Krebs cycle), thereby shifting the equilibrium to the right. The H$^+$ formation via this reaction is further increased by the activity of Cl$^-$/ HCO$_3^-$ exchanger (mainly AE2 (Stuart-Tilley et al., 1994), but also SLC26A7 (Petrovic et al., 2003)). These transport proteins operate in the basolateral membrane where they remove HCO$_3^-$ from the cell in exchange for Cl$^-$ (Muallem et al., 1985; Paradiso et al., 1987). This antiport has a dual function in the parietal cell: apart from its positive effect on cytosolic H$^+$ production, it also helps to replenish the cell with Cl$^-$ ions.

The importance of these processes becomes evident considering the consequences if any component in this H$^+$ supplying machinery is inhibited: Inhibition of either the carbonic anhydrase by azetazolamide or of the AE2 exchanger by DIDS[2] severely affects proton secretion of parietal cells. Moreover, AE2-deleted transgenic mice were shown to suffer from achlorhydria (Gawenis et al., 2004).

Osmotic balance

The continous secretion of osmotically active H$^+$ and Cl$^-$ ions into the stomach lumen leads to a substantial loss of parietal cell volume (Sonnentag et al., 2000). Although the majority of the water which is osmotically driven into the stomach lumen permeates through paracellular pathways, thus not directly depleting the parietal cell from water, it was shown that the basolateral membrane contains aquaporin-4 (Ma et al., 1994; Valenti et al., 1994; Misaka et al., 1996), thus allowing water molecules to leave the cell when it becomes osmotically less active. In contrast to many other cell types that regulate their volume by Na$^+$ K$^+$/2 Cl$^-$ cotransport via the NKCC symporter, the shrinkage of parietal cells is counteracted by the simultaneous activity of the basolaterally expressed Cl$^-$/HCO$_3^-$ and Na$^+$/H$^+$ (NHE) exchangers (Lang et al., 1998; Sonnentag et al., 2000). (NHE4 is highly expressed in parietal cells accoring to (Orlowski et al., 1992; Rossmann et al., 2001)). The combined activity of these two transporters is equivalent to a net uptake of NaCl into the cell. The resulting increase in

---

[2]4,4'-diisothiocyanostilbene-2,2'-disulfonic acid

intracellular Na$^+$ stimulates Na,K-ATPase activity, which in turn replenishes the cell with K$^+$ ions that are removed from the cell as a consequence of ongoing KCNQ1/KCNE2 activity. The importance of the osmotic balance mediated by these transporters is reflected by the hypochlorhydria phenotype of NHE4$^{-/-}$ mice. The reduction in acid secretion of these mice was accompanied by histological abnormalities in the gastric mucosa, such as an increased number of necrotic and apoptotic cells (Gawenis et al., 2005).

The transporter and channels that are involved in the overall process of gastric acid formation by the parietal cell are summarized in Figure 1.8 (inset).

1.2.2.3. Inhibitors of gastric H,K-ATPase

The gastric H,K-ATPase represents an important pharmacological target for the treatment of gastric acid-related diseases, including peptic ulcer disease (PUD) and gastro-esophageal reflux disease (GERD). Although the introduction of histamine-2 receptor antagonists (Black et al., 1972) (e.g. cimetidine, ranitidine) dramatically improved healing of PUD, this class of gastric acid inhibitors was less effective in the treatment of GERD which requires greater inhibition of gastric acid secretion. Moreover, since pharmacological tolerance to these medications develops in all patients (Lachman and Howden, 2000), also the efficacy of PUD treatment was reduced if the therapy continued over a prolonged period (maintenance therapy), which is usually necessary to prevent the ulcers from recurring. (*H. pylori* as a major pathogenic factor in the development of PUD was discovered years later (Marshall and Warren, 1984), followed by additional treatment of PUD with antibiotics to eradicate the bacteria.)

Sufficient suppression of gastric acid production for the succesful treatment of GERD was only achieved after the discovery of compounds targeting the final step of the secretion process, i.e. the H$^+$ transporting enzyme itself, which was identified in 1976 to be an electroneutral ATP-dependent H$^+$/K$^+$ exchanger (Sachs et al., 1976). Accordingly, these drugs inhibit acid secretion irrespective of the nature of the stimulus (either ligands acting via extracellullar receptors, e.g. histamine, gastrin, acetyl choline or intracellular second messenger, such as cAMP). Due to their different mechanism of action, these proton pump inhibitors are divided into the following two classes:

Proton Pump Inhibitors PPIs (covalent, irreversible Inhibitors)
The majority of PPIs are benzimidazole derivatives (e.g. omeprazole), but recently some promising new compounds of similar structure (imidazopyridine derivatives, e.g. tenatoprazole) were shown to have an enhanced plasma life time (Galmiche et al., 2004). Upon acidic activation of these inactive prodrugs to thioreactive species, they covalently attach to lumenally exposed cysteines of the α-subunit (see below). As listed in Table 1.3, the cysteine residues that are modified by the compounds differ for different PPIs (Shin et al., 1993; Besancon et al., 1993, 1997; Shin et al., 2006). However, all react with Cys-813 in the loop between TM5 and TM6, thereby permanently locking the the proton pump in the E$_2$ configuration. Figure 1.9 illustrates that the activation of the inactive prodrug involves two subsequently occurring acid-dependent steps (Lindberg et al., 1986):

i) Protonation of the pyridine moiety is responsible for the accumulation of PPIs in the acidic space of the stimulated parietal cell by converting the uncharged, membrane permeable drug into a positively charged, membrane impermeable compound. This results in 1000-fold higher concentrations of the drug at its site of action than in the blood.

ii) Protonation of the benzimidazole moiety results in formation of a transient cyclic intermediate which spontaneously rearranges to sulfenic acid, a highly thiophilic reagent. It can either directly react with cysteines or dehydrate to form sulfenamide, another thioreactive species that binds covalently to lumenally exposed thiols. It is crucial that this acid-catalyzed activation steps occurs as a direct consequence of the H$^+$ secreting activity of the H,K-ATPase itself, since the resulting highly reactive species will irreversibly bind to thiols in its immediate vicinity. Due to this acid activation mechanism of the prodrug, all PPIs must be given with acid protective coating to prevent conversion to the active compound already in the lumen of the stomach, where it would react with any available sulfhydryl group in food.

1.2. Physiological Importance of ATPases

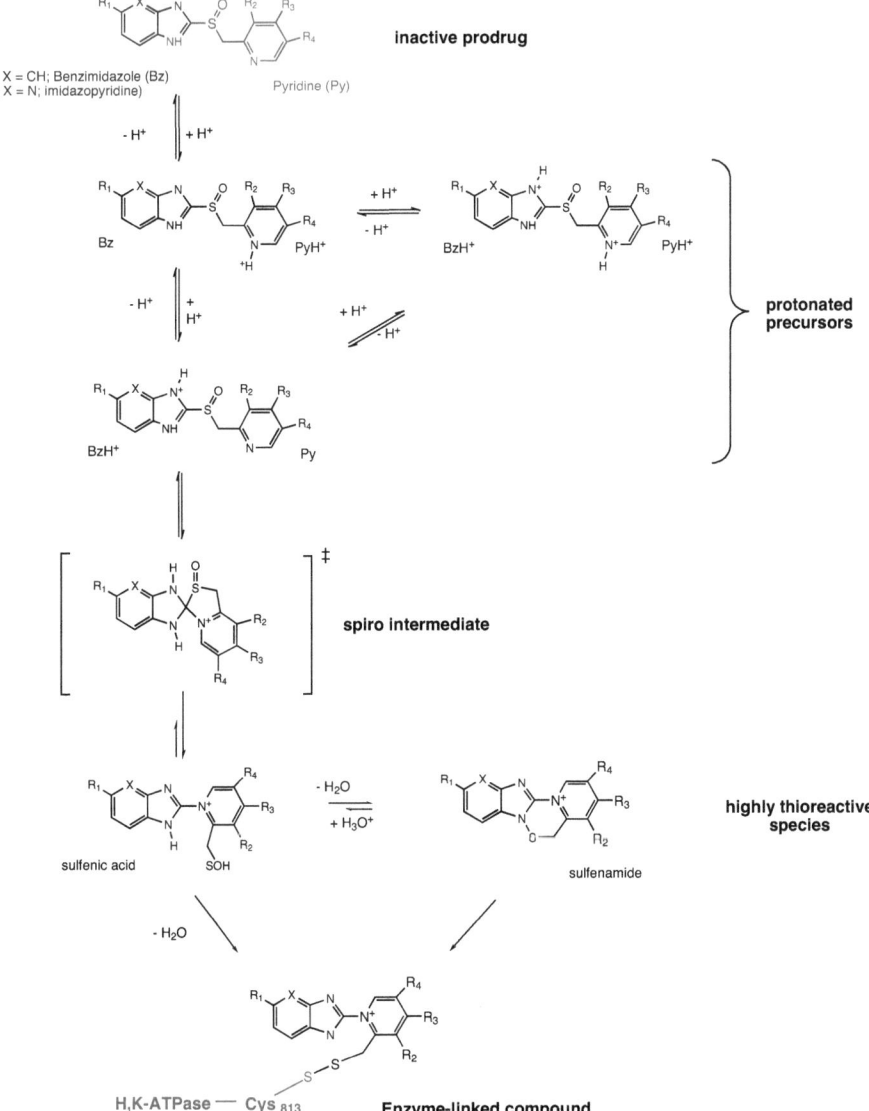

Figure 1.9.: **The mechanism of activation of PPIs.** (See Table 1.3 for substituents X, R1-R4)

PPI	Brand name	X	R1	R2	R3	R4	Site of reaction
Timoprazole	-	C	-	-	-	-	Cys813, Cys892
Omeprazole (racemate)	Antra MUPS® Gastracid®	C	OCH$_3$	CH$_3$	OCH$_3$	CH$_3$	Cys813
Esomeprazole (S-enantiomer)	Nexium®	C	OCH$_3$	CH$_3$	OCH$_3$	CH$_3$	Cys813
Lansoprazole	Agopton® Lanzor®,	C	-	CH$_3$	OCH$_2$CF$_3$	-	Cys813, Cys321
Rabeprazole	Pariet® Pariet Sieben®	C	-	CH$_3$	OCH$_2$CH$_2$CH$_2$OCH$_3$	-	Cys813, Cys321 Cys892
Pantoprazole	Pantozol® Rifun®	C	OCFH$_2$	OCH$_3$	OCH$_3$	-	Cys813, Cys822
Tenatoprazole	in Phase II clinical trials	N	OCH$_3$	CH$_3$	OCH$_3$	CH$_3$	Cys813, Cys822

Table 1.3.: **Different Proton Pump Inhibitors and their sites of reaction on the H,K-ATPase α-subunit.** X, R1, R2, R3 and R4 refer to the substituents on the core structure shown in Figure 1.9

The selective accumulation, the locally confined activation of the prodrug and the irreversibility of the covalent attachment to the proton pump contribute substantially to the long-lasting inhibition, which usually persists for 2-3 days after a single application. However, the requirement for acidic activation also leads to a lag-phase in the inhibition (Fellenius et al., 1981; Sachs et al., 2006; Sachs, 2003), representing a major disadvantage of this clinically most important class of inhibitors, which is especially relevant in the treatment of GERD.

Acid Pump Antagonists APAs (reversible Inhibitors)

In 1983 it was recognized that protonatable amines act as K$^+$-competitive and therefore reversible inhibitors of the H,K-ATPase (Im et al., 1984). This class of inhibitors is accordingly referred to as "**P-CABs**" (**P**otassium **c**ompetitive **a**cid in**hib**itors). The imidazopyridine compound SCH 28080 (see Figure 1.10 A) was developed as a homolog of omeprazole (Kaminski et al., 1985; Beil et al., 1986; Wallmark et al., 1987; Keeling et al., 1988). In contrast to PPIs, the inhibition by APAs does not rely on the H$^+$ secreting activity of the proton pump, thus resulting in a more rapid and complete inhibition of acid secretion. Yet, an acidic environment improves the efficacy of inhibition by SCH 28080 (pK$_a$ = 5.5), since only the protonated form can bind to the H,K-ATPase. Notably, as illustrated in Figure 1.10 B, SCH 28080 binds selectively to the E$_2$ or E$_2$P form of the H,K-ATPase (Wallmark et al., 1983; Shin and Sachs, 2004). Keeping in mind that omeprazole locks the enzyme in the E$_2$-state, it is interesting to note that inhibition by SCH 28080 and omeprazole is mutually exclusive, thus indicating an overlap in the binding region of the two compounds (Hersey et al., 1988).

Despite providing a very high affinity, superior inhibition properties and also effective inhibition of gastric acid secretion in humans (Ene et al., 1982), further development of many APAs including SCH 28080 was dropped as a result of liver toxicity in rats (Long et al., 1983). Only a few APA compounds having parent structures other than imidazo-pyridine or benzimidazoles are still being developed. Revaprazan, a pyrimidine derivative shown in Figure 1.10 A is now in phase II clinical trials in the United States and already clinically used in the Far East.

1.2. Physiological Importance of ATPases

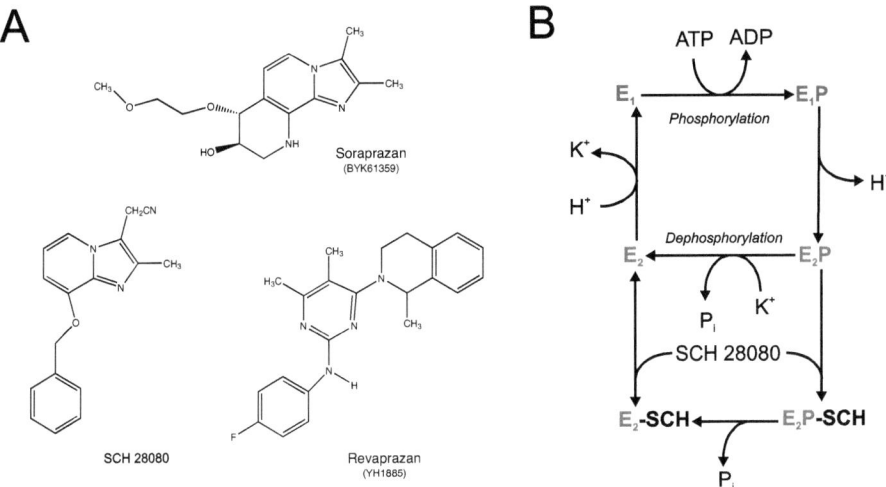

Figure 1.10.: **The structure of SCH 28080 and two other APAs (A) and a simplified Post-Albers-scheme illustrating the proposed inhibition mechanism for APAs (B).**

Aim and outline of this thesis

Na,K-ATPase and gastric H,K-ATPase are primary active transporters that belong to the superfamily of P-type ATPases. These membrane proteins utilize the energy released by the hydrolysis of ATP to transport their cationic substrates against an electrochemical gradient. Despite more than 50 years of extensive research on P-type ATPases, the molecular details of their ion transport mechanisms are not fully understood. Especially for oligomeric P-type ATPases, such as Na,K- and H,K-ATPases, which require an accessory β-subunit for their transport activity, many open questions remain, since only recently structural information became available (Morth et al., 2007; Abe et al., 2009) that revealed interesting particularities of these enzymes.

The β-subunits of this P-type ATPase sub-class (P_{2C}) are required for the correct folding, maturation and plasma membrane targeting of the catalytically active α-subunits. Furthermore, they have been shown to modulate the transport activity of the ion pumps. In contrast to the α-subunits of Na,K- and gastric H,K-ATPase which share about 60% sequence identity, there is less homology between their β-subunits (about 20-30% overall sequence identity between the gastric H,K-ATPase β-subunit and different Na,K-ATPase isoforms). Notably, however, they share a common basic structure: A short cytosolic N-terminal tail followed by a single transmembrane domain and a relatively large C-terminal ectodomain. The extracellular domain of almost all known β-subunits contains three highly conserved disulfide bridges and several N-glycosylation sites. Whereas the integrity of the disulfide bridges is essential for ATPase activity of both Na,K-and H,K-ATPase, the functional significance of the conserved N-linked oligosaccharides is less clear. A few conserved residues are also found in the transmembrane domain (TM). Of note, these conserved amino acids are all aromatic and located to one side of the helix, thus forming a patch for interactions with transmembrane domains of the respective α-subunit. Yet, the short N-terminal tail shows no homology at all between β-subunits of the two related ATPases and even exhibits a slight difference in length.

The aim of the research described in this thesis was to clarify how the β-subunits of the two aforementioned ATPases influence the cation transport activity of their respective α-subunit and whether common molecular mechanisms are involved in the modulation of ion transport properties by the β-subunits. Whereas already several mutational studies have demonstrated which parts of the Na,K-ATPase β-subunit may be responsible for its influence on ion transport activity of the catalytic Na,K-ATPase α-subunit, far less is known for the gastric H,K-ATPase β-subunit, since the ion transport properties of this electroneutrally operating transporter cannot be investigated by standard electrophysiological techniques. Voltage-clamp fluorometry, which combines two-electrode voltage-clamp and fluorescence measurements on site-specifically labeled H,K-ATPase and quantification of rubidium uptake by atomic absorption spectroscopy in *Xenopus* oocytes were therefore used to gain information about the conformational E_1P-E_2P distribution and the ion transport activity of the proton pump, respectively.

Using these biophysical techniques and biochemical methods as well, the functional consequences of mutational changes in all three topogenic domains of the β-subunit were investigated for the two ATPases : mutagenic changes in the conserved regions of β-TM (two tyrosines, see chapter 4) and ectodomain (N-glycosylation sites, see chapter 3) were systematically analyzed to clarify whether the structurally conserved β-regions are functionally important for ion transport activity. Furthermore, the potential interaction sites between β-TM and α-TM7 proposed by the two available Na,K-ATPase crystal structures (in the Rb^+-occluded E_2-state) were studied for both enzymes by characterizing a variety of α-TM7 mutants (see chapter 4). Notably, the recently published Cryo-EM structure of the gastric H,K-ATPase in the pseudo E_2P-state suggested an interaction between the β-N-terminus and the P-domain of the α-subunit which may stabilize the E_2P conformation by a ratchet-like mechanism. A similar interaction between β-N-terminus and P-domain of the Na,K-ATPase α-subunit was not described in the literature so far, since the Na,K-ATPase crystal structure is not well resolved around the cytosolic and extracellular parts of the β-subunit. Furthermore, mutational changes or moderate truncation of the β-N-terminus had no significant effects on cation transport of the sodium pump. To test the functional relevance of the short cytosolic β-N-terminus for the two closely related ATPases, N-terminally truncated β-variants were not only examined for the gastric H,K-ATPase, but also for the Na,K-ATPase (chapter 5).

This thesis encompasses a comprehensive analysis of Na,K- and H,K-ATPase mutations in conserved β-subunit regions and other domains that may be relevant for interactions to the respective α-subunit according to the recently available structural data. Therefore, this work contributes to the understanding of how β-subunits influence the transport activity of P_{2C}-type ATPases and provides a clue to the question whether similar molecular mechanisms are responsible for this kind of cation transport modulation of the two related ion pumps.

CHAPTER 2

Material and Methods

2.1. Molecular biology

2.1.1. Expression vectors and cDNA-constructs

The cDNAs of the sheep Na,K-ATPase β_1-subunit, rat H,K-ATPase β-subunit, rat gastric H,K-ATPase α-subunit and a modified form of the sheep Na,K-ATPase α_1-subunit without extracellularly exposed cysteine residues (containing mutations C911S and C964A (Hu and Kaplan, 2000)) and with reduced ouabain sensitivity in the millimolar range (achieved by the mutations Q111R and N122D (Price and Lingrel, 1988)) were subcloned into vector pTLN (Lorenz et al., 1996). The reduced ouabain sensitivity of the latter construct allows selective inhibition of the endogenous *Xenopus* Na,K-ATPase and was therefore used for all coexpression studies with mutated β_1-constructs or as template for site-directed mutagenesis of the Na,K-ATPase α-subunit. Furthermore, to exclude any background signals in voltage-clamp fluorometric studies, which could possibly arise from Na,K-ATPase enzymes assembled from heterologously expressed α-subunits and endogenous β-subunits, we utilized the β_1-subunit sequence variant S62C for mutagenesis. The introduced cysteine is close to the transmembrane/extracellular interface and has been shown to give rise to voltage-dependent fluorescence changes upon site-directed fluorescence labeling with TMRM, without impairing enzyme function (Dempski et al., 2005). To enable voltage-clamp fluorometry on gastric H,K-ATPase, we used a modified H,K-ATPase α-subunit with a single cysteine replacement of a serine in the M5/M6 extracellular loop (S806C), which is homologous to the N790C mutation of the Na,K-ATPase α-subunit (Geibel et al., 2003b) and thus also suited for environmentally sensitive TMRM-labeling (Geibel et al., 2003a). Rubidium uptake measurements confirmed that the S806C mutation did not affect the H,K-ATPase's transport properties (see Figure 2.1). This construct was therefore used as a template for site-directed mutagenesis to introduce mutations in the H,K-ATPase α-subunit or for coexpression with mutated H,K-ATPase β-subunits.

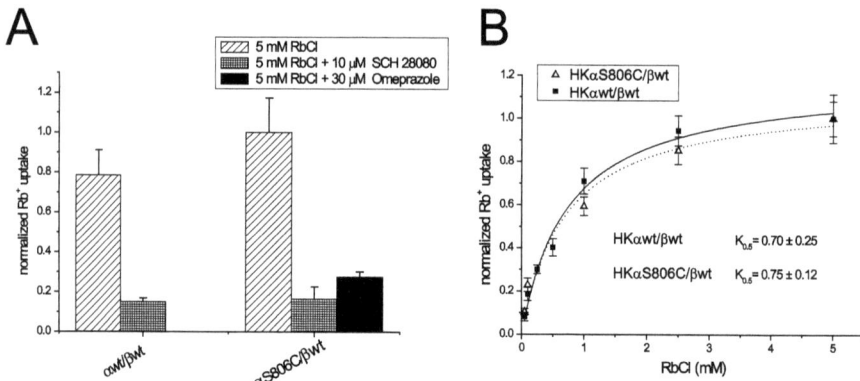

Figure 2.1.: **Rb^+ transport properties of the H,K-ATPase reporter construct HKαS806C.** A, H,K-ATPase-mediated Rb^+ uptake at 5 mM RbCl in the absence (hatched bars) or presence of 10 μM SCH 28080 (crossed bars) or 30 μM omeprazole (black bar). B, Michaelis-Menten-plots for Rb^+ uptake by H,K-ATPase, shown for oocytes expressing either HKαwt + HKβwt (filled squares) or HKαS806C + HKβwt (open triangles), respectively. Data are means ± S.E. (n = 8-12 oocytes) from 2-3 independent experiments, normalized to Rb^+ uptake at 5 mM RbCl, corresponding to values between 15-20 pmol/(oocyte·min).

2.1.2. Site-directed mutagenesis

All single point mutations were introduced using the Quikchange XL Site-Directed Mutagenesis Kit (Stratagene). Since the Quikchange Multi Site-Directed Mutagenesis Kit (Stratagene) enables the simultaneous mutation of multiple sites in a single PCR-step procedure, it was used to prepare glycosylation-deficient Na,K- and H,K-ATPase β-subunit variants (termed NaKβS62Cgd or HKβgd). The N-terminally truncated H,K-ATPase β-subunit mutants (termed HKβΔ4, HKβΔ8, HKβΔ13 and HKβΔ29) were constructed by deletion of 3, 7, 12 and 28 amino acids after the first methionine of the rat H,K-ATPase β-wildtype sequence, respectively. The deletion was achieved by a PCR using an antisense oligonucleotide recognizing the *Xenopus* β-globin 3' untranslated region and a sense oligonucleotide which contained a *BamH*I restriction site covering the Kozak sequence (Kozak, 1987) including the initiator methionine. This permitted cloning of the mutant cDNA into *BamH*I-containing wild-type pTLNβHK with *BamH*I and *Xho*I. The ORF sequence of all mutants was subsequently confirmed by DNA sequencing (MWG Eurofins Operon).

2.2. Oocyte preparation and cRNA injection

Xenopus oocytes were obtained by collagenase treatment after partial ovariectomy from *Xenopus laevis* females. cRNAs were prepared using the SP6 mMessage mMachine Kit (Ambion, Austin, Texas) and checked for potential degradation by gel electrophoresis. A 50 nl aliquot containing 20-25 ng of Na,K-ATPase α_1-subunit and 1.5-2.5 ng of Na,K-ATPase β_1-subunit cRNA (or 20-25 ng H,K-ATPase α-subunit cRNA and 5 ng of H,K-ATPase β-subunit cRNA) were injected into each cell. After injection, oocytes were kept in ORI buffer (110 mM NaCl, 5 mM KCl, 2 mM $CaCl_2$, 5 mM HEPES, pH 7.4) containing 50 mg/l gentamycin at 18 °C for 3 to 5 days for the Na,K-ATPase and for 2 days for the gastric H,K-ATPase respectively.

2.3. Membrane preparation from *Xenopus laevis* oocytes

2.3.1. Isolation of plasma membranes

The isolation of plasma membranes was carried out using positively charged silica beads as described by Kamsteeg and Deen (Kamsteeg and Deen, 2000), who reported a 25- or 450-fold higher yield with this technique compared to standard isolation or biotinylation procedures. Morever, since binding of these beads is carried out on intact oocytes, a high purity of the plasma membrane fraction is achieved with only negligible contaminations by internal membranes (Kamsteeg and Deen, 2001). After removal of their follicular cell layer, 8-12 oocytes were rotated in 1 % colloidal silica, (Ludox Cl, Sigma-Aldrich) in MES buffered saline for silica (MBSS; 20 mM MES, 80 mM NaCl, pH 6.0) for 30 min at 4 °C. After washing two times in MBSS, the oocytes were rotated at 4 °C in 0.1 % polyacrylic acid (Sigma-Aldrich) in MBSS for 30 min. This blocking agent is added to coat unbound beads and solvent-exposed surfaces of beads that were already bound to oocytes. Afterwards, the oocytes were washed two times in Modified Barth's Solution (MBS; 0.33 mM $Ca(NO_3)_2$, 0.41 mM $CaCl_2$, 88 mM NaCl, 1 mM KCl, 2.4 mM $NaHCO_3$, 0.82 mM $MgSO_4$, 10 mM HEPES, pH 7.5). Subsequently, oocytes were homogenized in 1.5 ml buffer HbA (20 mM TRIS, 5 mM $MgCl_2$ 5 mM NaH_2PO_4, 1 mM EDTA, 80 mM sucrose, pH 7.4) containing protease inhibitor (Complete; Roche Mol. Biochem., Mannheim, Germany) and centrifuged for 30 s at 10 g at 4 °C, after which 1.3 ml of the sample supernatant was removed (to be saved for subsequent preparation of total membranes, see next section 2.3.2) and 1 ml HbA was added to the silica beads. This centrifugation and exchange of HbA was repeated three times, but centrifugation changed from twice at 10 g via once at 20 g to once at 40 g. After the last centrifugation step, HbA was removed and plasma membranes were spun down for 30 min at 16,000 g at 4 °C and resuspended in Laemmli buffer (Laemmli, 1970) (4 μl/oocyte).

2.3.2. Preparation of total membranes from *Xenopus laevis* oocytes

To remove yolk platelets, the supernatant from homogenized oocytes in HbA containing protease inhibitor was centrifuged once for 3 min at 2,000 g at 4 °C, and the pellet discarded. Total membranes in the supernatant were spun down for 30 min at 16,000 g at 4 °C and resuspended in 4 µl/oocyte of Laemmli buffer (Laemmli, 1970).

2.3.3. Immunoblotting (Western Blot)

Protein samples equivalent to two oocytes were separated on 10-15% SDS polyacrylamide gels and blotted on nitrocellulose membranes (Roth). The α- and β-subunits of the sheep Na,K-ATPase were detected with the polyclonal anti-Na,K α-antibody C356-M09 (Koenderink et al., 2003) and the monoclonal anti-Na,K β-antibody M17-P5-F11 (Ball et al., 1992). The α- and β-subunits of the rat gastric H,K-ATPase were detected with the polyclonal anti-H,K α-antibody HK12.18 (Gottardi and Caplan, 1993) (Merck) and the monoclonal anti-H,K β-antibodies 2G11 (Chow and Forte, 1993)(Acris Antibodies) or 2B6 (Mori et al., 1989) (BIOZOL) respectively. Subsequently, blots were incubated with appropriate HRP-conjugated secondary antibodies (Dako) and proteins were visualized by using an enhanced chemiluminescence kit (Roche Molecular Biochemicals, Mannheim, Germany).

2.4. Rb^+ uptake measurements

2.4.1. Rb^+ uptake measurements

Two days after injection, noninjected control oocytes and H,K-ATPase expressing oocytes were preincubated for 15 min in a Rb^+- and K^+-free solution (90 mM TMACl or NaCl, 20 mM TEACl, 5 mM $BaCl_2$, 5 mM $NiCl_2$, 10 mM HEPES, pH 7.4) containing 100 µM ouabain to ensure inhibition of the endogenous Na,K-ATPase and then incubated for 15 min under temperature control in Rb^+-flux-buffer at 21 °C (5 mM RbCl, 85 mM TMACl or NaCl, 20 mM TEACl, 5 mM $BaCl_2$, 5 mM $NiCl_2$, 10 mM MES, pH 5.5, 100 µM ouabain). After 3 washing steps in Rb^+-free washing-buffer (90 mM TMACl or NaCl, 20 mM TEACl, 5 mM $BaCl_2$, 5 mM $NiCl_2$, 10 mM MES, pH 5.5) and one wash in water, each individual oocyte was homogenized in 1 ml of Milli-Q® water (Millipore, Billerica, MA). To determine the apparent constant $K_{0.5}$ for half-maximal activation of the H,K-ATPase by rubidium, the sum [TMACl or NaCl] plus [RbCl] in the Rb^+-flux buffer was kept constant at 90 mM, e.g. 1 mM RbCl + 89 mM TMACl (or NaCl). After subtraction of the mean of Rb^+ uptake into control oocytes of the same batch at a given RbCl concentration, the data were fitted to a Michaelis-Menten type function:

$$v = v_{max} \cdot \frac{[S]}{K_{0.5} + [S]} \quad (2.1)$$

where [S] represents the rubidium chloride concentration [RbCl], v_{max} the Rb^+ uptake at saturating $[Rb^+]$ and $K_{0.5}$ the apparent constant of halfmaximal activation.

2.4.2. Atomic Absorption spectroscopy (AAS)

Oocyte homogenates were analyzed by atomic absorption spectroscopy using an AAnalyst800™ spectrometer (Perkin Elmer, Waltham, MA). From oocyte homogenates (typically 1 ml) samples of 20 µl were automatically transferred into a transversely heated graphite furnace (THGF) and subjected to the temperature protocol shown in Table 2.1.

Absorption was measured at 780 nm during atomization using a Rubidium hollow cathode lamp (Photron, Melbourne, Australia). After Zeeman background correction, Rb^+ contents were calculated by comparison with standard calibration curves (measured between 0 and 50 µg/l Rb^+). The detection limit (characteristic mass[1]) of Rb^+ is ~10 pg.

---

[1] The characteristic mass is the quantity of analyte in picograms (pg) that would yield 1 % absorption, which is 0.0044 OD.

Step	Temperature (°C)	Ramp Time (sec)	Hold Time (sec)	Argon Gas Flow (ml/min)
Dry 1	110	1	20	250
Dry 2	130	5	30	250
Pyrolysis	300-600	10	20	250
Atomization	1700-1800	0	5	0
Clean-out	2400	1	2	250

Table 2.1.: **Temperature protocol for THGF-AAS.**

2.5. Electrophysiology

2.5.1. Oocyte pretreatment, fluorescence labeling and experimental solutions

Prior to functional studies on Na,K-ATPase-expressing oocytes, cells were first incubated for 45 min in Na^+-loading buffer (110 mM NaCl, 2.5 mM Nacitrate, 5 mM MOPS, 5 mM TRIS, pH 7.4), then for 15 min in Post-loading buffer (100 mM NaCl, 1 mM $CaCl_2$, 5 mM $BaCl_2$, 5 mM $NiCl_2$, 5 mM MOPS/TRIS, pH 7.4) to elevate the intracellular Na^+ concentration. For voltage-clamp fluorometry, site-specific labeling was achieved by incubating oocytes in Post-loading buffer containing 5 µM tetramethylrhodamine-6-maleimide (TMRM, Molecular Probes, stock solution 5 mM in DMSO) for 5 min at room temperature in the dark, followed by extensive washes in dye-free Post-loading buffer. Measurements under high extracellular Na^+/K^+- free conditions were carried out in Na^+-test-solution (100 mM NaCl, 5 mM $BaCl_2$, 5 mM $NiCl_2$, 5 mM MOPS/TRIS, pH 7.4, 10 µM ouabain). The following solution was used for VCF measurements on gastric H,K-ATPase: 90 mM TMACl or NaCl, 20 mM TEACl, 5 mM $BaCl_2$, 5 mM $NiCl_2$, 10 mM MES, pH 5.5 or pH 7.4.

2.5.2. Voltage-clamp fluorometry

An oocyte perfusion chamber was mounted in an Axioskop 2FS epifluorescence microscope (Carl Zeiss, Göttingen, Germany) equipped with a 40x water immersion objective (numerical aperture = 0.8). Currents were measured using a two-electrode voltage-clamp amplifier (Turbotec 05, npi, Tamm, Germany). Fluorescence was excited with a 100 W tungsten lamp using filters 535DF50 (excitation), 565 EFLP (emission) and 570DRLP (dichroic, all Omega Optical, Brattleboro, VT). Fluorescence was measured with a PIN-022A photodiode (UDT, Hawthorne, CA) mounted to the microscope camera port. Photocurrents were amplified by a low-noise current amplifier DLPCA-200 (FEMTO, Berlin, Germany). Fluorescence and currents were recorded simultaneously using a Digidata 1322A interface and subsequently analyzed with Clampex 9.2 and Clampfit 9.2 software (Molecular Devices, Sunnyvale, CA).

2.5.3. Analysis of stationary currents of the Na,K-ATPase

Stationary currents of the Na,K-ATPase were measured upon a solution exchange from 0 mM to 10 mM K^+ (10 mM KCl, 90 mM NaCl, 5 mM $BaCl_2$, 5 mM $NiCl_2$, 5 mM MOPS/TRIS, pH 7.4, 10 µM ouabain). To determine the apparent constant $K_{0.5}$ for half-maximal activation of the Na,K-ATPase by K^+, the sum [NaCl] plus [KCl] in the Na-Test-solution was kept constant at 100 mM, e.g. 1 mM KCl + 99 mM NaCl. After run-down correction of stationary currents (if necessary) at a given K^+ concentration, the data were fitted to a Michaelis-Menten type function (equation 2.1). The turnover numbers of wildtype or mutant Na,K-ATPase pumps were calculated by dividing the stationary current (at 10 mM K^+ and -40 mV) by the total moved charge $Q_{tot} = Q_{max} - Q_{min}$ (from fits of Q-V curves with a Boltzmann function, see section 2.5.4), both measured on the same oocyte.

2.5.4. Analysis of transient currents of the Na,K-ATPase

Pre-steady state currents under high extracellular Na^+/K^+-free conditions were obtained by subtracting the current responses to voltage steps from -40 mV to values between +60 mV and -180 mV

(20 mV increments) in presence of 10 mM ouabain (inhibiting the endogenously as well as the heterologously expressed pump) from currents measured in presence of 10 µM ouabain (inhibiting only the endogenous Na,K-ATPase). The resulting difference currents were fitted with a single-exponential function, disregarding the first 3-5 milliseconds after the voltage step to exclude capacitive artifacts. The translocated charge Q was calculated from the integral of the fitted transient currents, and the resulting Q-V curves were approximated by a Boltzmann-type function

$$Q(V) = Q_{max} + \frac{Q_{max} - Q_{min}}{(1 + \exp(\frac{z_q \cdot F}{R \cdot T}(V - V_{0.5})))} \quad (2.2)$$

where Q_{min} and Q_{max} are the saturating values of translocated charge, $V_{0.5}$ is the midpoint potential, z_q the fraction of charge displaced through the entire transmembrane field, F the Faraday constant, R the molar gas constant, T the temperature (in K) and V the transmembrane potential. All experiments were performed at 22- 24 °C.

2.6. Extracellular pH measurements

A qualitative assay for the acidification of the extracellular medium mediated by H^+ secretion of *Xenopus* oocytes expressing gastric H,K-ATPase was carried out according to Jaisser et al. (Jaisser et al., 1993). Two days after injection, oocytes were incubated for 5 min in a weakly buffered solution (70 mM TMACl, 20 mM RbCl, 5 mM $BaCl_2$, 5 mM $NiCl_2$, 500 µM MOPS, adjusted to pH 7.4 with TMAOH), containing the pH indicator phenol red (200 µg/ml). In some experiments, SCH 28080 (100 µM) was added to the solution. Oocytes were placed individually in a small drop (0.5-1 µl) of the same solution under mineral oil. Every two minutes a color picture was taken at room temperature.

CHAPTER 3

Characterization of Na,K-ATPase and H,K-ATPase Enzymes with Glycosylation-Deficient β-Subunit Variants by Voltage-Clamp Fluorometry in *Xenopus* Oocytes

Dürr et al. (2008), Biochemistry 47, 4288–4297

3.1. Introduction

A common feature of the H,K-ATPase β-subunit and all Na,K-ATPase β-isoforms is the presence of several conserved N-glycosylation sites on the C-terminal extracellular domain (illustrated in Figure 3.1), resulting in a substantial increase of the mature subunit's molecular weight. This increase ranges from at least 80% for the Na,K-ATPase $β_1$-subunit, having only three N-linked glycosylation sites, up to more than 100% for the Na,K-ATPase $β_2$-subunit with 4 to 9, or the H,K-ATPase β-subunit possessing 6 to 7 glycosylation sites. Although studied in various expression systems, questions regarding the functional significance of these huge carbohydrate moieties have not been completely settled yet. Whereas removal of N-linked glycosylation on H,K-ATPase β-subunits had dramatic consequences in terms of α/β coassembly (Asano et al., 2000), plasma membrane delivery (Asano et al., 2000; Vagin et al., 2003) and catalytic activity ((Asano et al., 2000; Klaassen et al., 1997) of the holoenzyme, no such effects have been observed for the deglycosylated Na,K-ATPase (Takeda et al., 1988; Tamkun and Fambrough, 1986; Sun and Ball, 1994; Beggah et al., 1997; Zamofing et al., 1989). However, the N-linked carbohydrates on the Na,K-ATPase β-subunit are crucial for some lectin-like properties (Kitamura et al., 2005), including a role in cell-cell adhesion of epithelial cells (Vagin et al., 2006), neuronal interactions of $β_2$/AMOG1-expressing glial cells (Heller et al., 2003) and isoform-specific basolateral versus apical targeting in various polarized cells (Vagin et al., 2005a; Lian et al., 2006; Vagin et al., 2007). Likewise, the N-glycans on the H,K-ATPase have been shown to be essential for apical targeting (Vagin et al., 2007, 2004), protection from proteolytic digestion (Thangarajah et al., 2002; Crothers et al., 2004) and acidic denaturation (Tyagarajan et al., 1996).

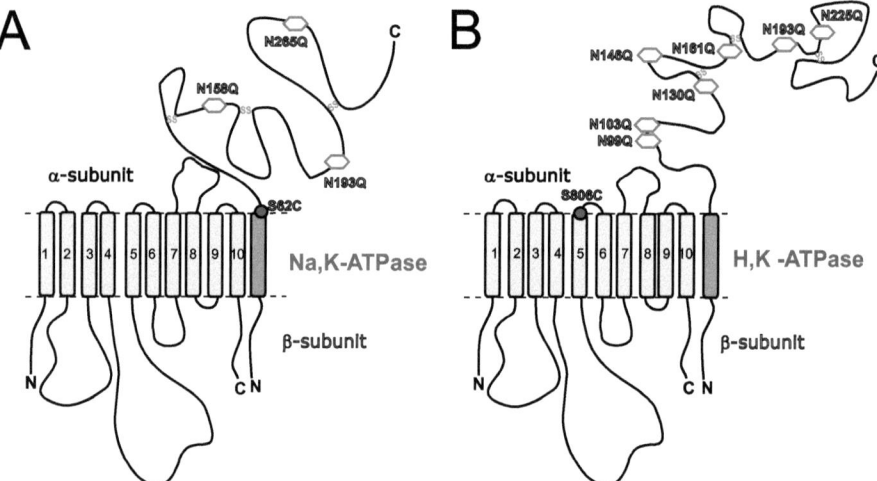

Figure 3.1.: **N-glycosylation sites and cysteine mutations for site-specific labeling of Na,K- (A) and H,K-ATPase (B).** Na,K-ATPase (A): the reporter site Ser-62C and the three mutated N-glycosylation sites (hexagons) on the Na,K-ATPase β-subunit are shown. H,K-ATPase (B): the reporter site Ser-806C on the H,K-ATPase α-subunit and the seven mutated N-glycosylation sites (hexagons) on the H,K-ATPase β-subunit (dark grey) are shown. "S-S" denotes disulfide bridges, amino- and carboxy-termini are indicated as N and C, respectively.

Thus, to differentiate between mainly *in situ* relevant secondary effects and primary effects on the functional properties of the enzyme itself, we aimed to investigate the as yet unknown functional significance of N-linked glycans for the H,K-ATPase using the *Xenopus* oocyte expression system. Another aspect of N-linked glycosylation, which has not been examined in P-type ATPases so far,

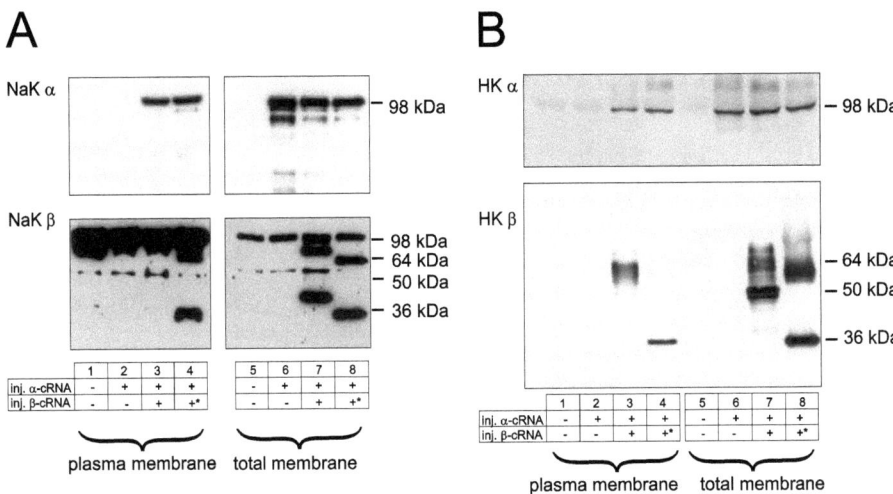

Figure 3.2.: **Western blot analysis of plasma membrane and total membrane fractions from Na,K- and H,K-ATPase-expressing oocytes.**
(A) Plasma membrane and total membrane fractions from oocytes injected with cRNAs for: NaKα only (lanes 2 and 6), NaKα + NaKβS62C (lanes 3 and 7), or NaKα + NaKβS62Cgd (lanes 4 and 8, see asterisks in the injection scheme), or from uninjected oocytes (lanes 1 and 5). Detection used Na,Kα-specific antibody C356-M09 (upper panel) or Na,K-specific antibody M17-P5-F11 (lower panel). (B) Plasma membrane and total membrane fractions from oocytes injected with cRNAs: for HKαS806C only (lanes 2 and 6), HKαS806C + HKβwt (lanes 3 and 7), or HKαS806C + HKβgd (lanes 4 and 8, see asterisks in the injection scheme), or from uninjected oocytes (lanes 1 and 5). Detection used anti-H,Kα antibody HK12.18 (upper panel) or anti-H,Kβ antibody 2G11 (lower panel).

is whether the presence of the huge oligosaccharide moiety has any effect on the kinetics of the $E_1P \rightarrow E_2P$ conformational transition or the distribution between E_1P/E_2P states of the catalytic cycle. Remarkably, the magnitude of fluorescence intensity changes upon K^+ addition to FITC-labeled, purified Na,K-ATPase was about 50% larger compared to the fully glycosylated enzyme, when N-linked oligosaccharides had been removed by a combined neuraminidase/endo-F treatment of the holoenzyme (Sun and Ball, 1994). This observation possibly reflects a shift in the enzymes' distribution between E_1P/E_2P states. Regarding the substantial effect of N-linked sugars on the molecular weight of the holoenzyme, such a shift could arise from a kinetic effect on some particular steps in the catalytic cycle, which may in turn lead to changes in the apparent cation affinities.
To test this hypothesis, we utilized the technique of voltage-clamp fluorometry, which is the only available technique that allows presteady-state kinetic investigations of electroneutrally operating ion transporters like the gastric H,K-ATPase in intact cells. This enabled us to directly monitor the voltage-dependent distribution of fluorescence labeled H,K- or Na,K-ATPases between E_1P/E_2P states of the catalytic cycle in a time-resolved fashion.

3.2. Results

3.2.1. Plasma Membrane Delivery and α-Subunit Stabilization of Glycosylation-Deficient Mutants

To assess the question whether N-linked glycosylation is necessary for cell surface delivery of the Na,K- or H,K-ATPase, plasma membranes were isolated from injected oocytes and compared to total membrane preparations by Western blotting. Importantly, the glycosylation-deficient variants of both Na,K- and H,K-ATPase were delivered to the plasma membrane (lanes 4 in Figure 3.2 A and B, respectively). Regarding the H,K-ATPase wildtype β-subunit, the broad region from 60 to 70 kDa detected by the anti-H,K β-antibody (Figure 3.2 B, lane 3) is characteristic for complex-type oligosaccharides due to considerable variety in number and composition of terminal sugars, giving rise to the heterogeneity of the mature glycosylated H,K-ATPase β-subunit. Likewise, a similar broad band would be expected for the fully glycosylated Na,K β-subunit in the respective plasma membrane fraction, but only a rather weak discrete band is stained by the anti-Na,K β-antibody. Since it is also present in all other lanes from Figure 3.2 A, including control preparations from uninjected or only α-subunit cRNA injected oocytes, it must be mainly attributed to the endogenously expressed *Xenopus laevis* Na,K-ATPase β-subunit. Although the α-subunit is strongly expressed, a substantially more intense band indicating expression of heterologous Na,K-ATPase β-subunit is absent in the corresponding plasma membrane fraction (Figure 3.2 A, lane 3). Notably, the epitope recognized by the M17-P5-F11 antibody is only two amino acids separated from the second glycosylation site N193 (Sun and Ball, 1994). Obviously, the majority of the various complex-type oligosaccharides attached to this site impairs antibody binding, leading to a low detection efficiency for the glycosylated protein. This interpretation is in agreement with previous studies which have demonstrated a 10-fold stronger antibody binding to this epitope after enzymatic removal of N-linked carbohydrates (Sun and Ball, 1994). Moreover, this is supported by the intense focused band at 41 kDa observed in the total cellular membrane fraction (Figure 3.2 A, lane 7) prepared from the same oocyte sample as the plasma membrane preparation, which is characteristic for Na,K-ATPase β-subunits carrying high-mannose precursor oligosaccharides. Since these are readily detected by the antibody, apparently the terminal sugars added in the very last steps of glycosylation interfere with antibody binding. As these high mannose precursors were reported to remain intracellularly, being unable to reach the plasma membrane until completely glycosylated (Takeda et al., 1988; Tamkun and Fambrough, 1986), this band is exclusively observed in the total cellular membrane fraction. Moreover, its absence in the plasma membrane fraction excludes any contamination of this sample by total cellular membranes. Analogously, the 50 kDa band uniquely present in the HKαwt/HKβwt total cellular membrane fraction (Figure 3.2 B, lane 7) is characteristic for the H,K-ATPase β-subunit's immature, mannose-rich glycosylation.

For the glycosylation-deficient Na,K- and H,K-ATPase β-mutants, these bands are each shifted to lower molecular weights of about 32–34 kDa for both enzymes, corresponding to the protein core, which is seen in both plasma- (Figure 3.2 A and B, lane 4) and total membrane fractions (Figure 3.2 A and B, lane 8). The band at high molecular weights in the total membrane fraction of glycosylation-deficient H,K-ATPase (Figure 3.2 B, lane 8) could potentially represent aggregates formed between the fully deglycosylated H,K-ATPase β-subunits, which do not dissociate even under denaturing conditions. It has been proposed that core sugars have a global effect on the solubility of newly synthesized proteins that counteracts the tendency to form irreversible aggregates by hydrophobic interactions (Marquardt and Helenius, 1992).

Notably, the glycosylation-deficient Na,K- and H,K-ATPase β-subunits were able to stabilize their corresponding α-subunits equally well as their glycosylated counterparts, as judged from the amount of the 100 kDa α-subunit protein detected in the plasma membrane fraction (Figure 3.2 A and B, compare upper lanes 3 and 4 each). In contrast, when Na,K- or H,K-ATPase α-subunits are expressed without their corresponding β-subunits, the heterologous cRNA is translated into protein, which, however, is not targeted to the plasma membrane, but remains in intracellular (mainly ER) membranes, thus explaining the presence of a 100 kDa band in both lanes 6 and its absence in both lanes 2 in Figure 3.2 A and B. Furthermore, the intracellularly retained Na,K-ATPase α-subunit is more susceptible to cellular degradation, as indicated by the bands below 100 kDa also detected by the Na,K α-antibody in the total membrane preparation (Figure 3.2 A, lane 6). This phenomenon has been reported previously (Geering et al., 1989). The results from these control experiments support the notion that the observed plasma membrane stabilization of the α-subunits is indeed brought about by association with the coexpressed glycosylation-deficient β-subunits, rather than by assembly with the endogenously expressed Na,K-ATPase β-subunits.

3.2.2. Enzyme Activity of Glycosylation-Deficient Mutants

To test whether the enzymes assembled from glycosylation-deficient Na,K- and H,K-ATPase β-subunit mutants are still functional in the plasma membranes of *Xenopus* oocytes, the transport activity was assessed by measuring K^+ activated stationary currents for the electrogenic Na,K-ATPase and by rubidium uptake measurements for the electroneutrally operating H,K-ATPase. Table 3.1 shows the stationary currents observed under saturating K^+ concentrations. Values are identical within error limits for the glycosylation-deficient and the glycosylation-competent enzyme. Similarly, the turnover number is not influenced by the glycosylation status (see Table 3.1). Thus, both the amount of functional enzyme on the cell surface and ion pumping activity of the Na,K-ATPase are not affected by the removal of N-linked oligosaccharides from the β-subunit.

As illustrated in Figure 3.3 A, Rb^+ transport activity under saturating extracellular rubidium concentrations is also quite similar when comparing oocytes expressing glycosylated and glycosylation-deficient H,K-ATPase complexes, and is clearly higher than in uninjected oocytes or in cells expressing only H,K-ATPase α-subunits. Specificity of Rb^+ uptake is demonstrated by inhibition with the H,K-ATPase inhibitor SCH 28080 (Figure 3.3 A). Furthermore, the apparent constant for half-maximal Rb^+ activation ($K_{0.5}$) derived from Figure 3.3 B (inset) is very similar for glycosylated and glycosylation-defective H,K-ATPase (0.67 ± 0.04 mM versus 0.64 ± 0.04 mM).

construct	stationary current (nA)	turnover number (s^{-1})
Na,Kα_1WT/β_1S62C	255 ± 24	34.2 ± 3.2
Na,Kα_1WT/β_1S62Cgd	205 ± 19	32.5 ± 2.3

Table 3.1.: **Stationary currents and turnover number of Na,K-ATPase α/β complexes containing wildtype or non-glycosylated β-subunits.** Turnover numbers were determined as described in section 2.5.3. Values are means ± S.E. of 5-12 oocytes from two different oocyte batches.

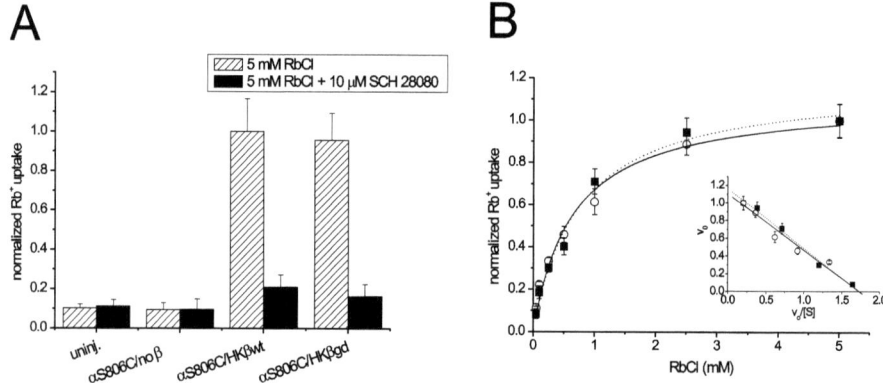

Figure 3.3.: **Rb$^+$ uptake into oocytes by wildtype and glycosylation-deficient H,K-ATPase enzymes.** A, H,K-ATPase-mediated Rb$^+$ uptake at 5 mM RbCl in the absence (hatched bars) or presence (black bars) of 10 μM SCH 28080. Results from uninjected control oocytes, or oocytes injected with the following cRNAs are shown: HKαS806C only, HKαS806C + HKβwt, or HKαS806C + HKβgd. Data in each experiment were normalized to H,KαS806C/βwt H,K-ATPase Rb$^+$ uptake in absence of SCH 28080, which were 14.0, 18.6 and 19.7 pmol/(oocyte·min), respectively. B, Michaelis-Menten-plot for Rb$^+$ uptake by H,K-ATPase. Inset: Eadie-Hofstee-plot. Oocytes were injected with HKαS806C- and HKβwt-cRNA (filled squares) or with HKαS806C- and HKβgd-cRNA (open circles), respectively. Data in each experiment were normalized to Rb$^+$ uptake at 5 mM RbCl, corresponding to 18.5 and 9.9 pmol/(oocyte·min) for HKαS806C/βwt; 14.1, 20.7 and 18.7 pmol/(oocyte·min) for HKαS806C/βgd. Data are means ± S.E., n = 8-10 oocytes from 2 or 3 experiments.

3.2.3. E_1P/E_2P Conformational Distribution and Kinetics of $E_1P \rightarrow E_2P$ Transition of Glycosylation-Deficient Mutants

To investigate whether removal of the huge sugar moiety influences the kinetics of the $E_1P \rightarrow E_2P$ conformational change, the mutated ATPases were also studied by voltage-clamp fluorometry. TMRM-labeled oocytes expressing the wildtype Na,K-ATPase α-subunit together with either glycosylation-competent S62C or glycosylation-deficient S62Cgd β-subunits were subjected to voltage pulses and the resulting conformation-dependent fluorescence changes (Figure 3.4, A) were analyzed. When a Boltzmann function is fitted to the corresponding (1-ΔF/F)-V distributions, the voltage dependence is very similar for glycosylated and glycosylation-deficient enzymes (Figure 3.5 A), thus yielding identical fit parameters within error limits (Table 3.2). As illustrated in Figure 3.5 B, these (1-ΔF/F)-V curves are strictly correlated to the voltage-dependent distribution of charge movement (Q-V curves, see Table 3.2 for Boltzmann parameters), which was determined from transient currents recorded in parallel (Figure 3.4 B).

Time constants for the voltage-dependent $E_1P \rightarrow E_2P$ transition resulting from monoexponential fits of the fluorescence signals are presented in Figure 3.5 C for S62C and the glycosylation-deficient S62Cgd variant, respectively. Likewise, Figure 3.5 D shows the reciprocal time constants of the simultaneously recorded transient currents, which are about 4 times larger than those from fluorescence changes at hyperpolarizing potentials, as observed previously (Dempski et al., 2005). Since both time constants determined for the glycosylation-deficient mutant are comparable to the fully glycosylated enzyme, the kinetics of the voltage-dependent $E_1P \rightarrow E_2P$ transition seems unaffected by the bulky sugars on the β-subunit.

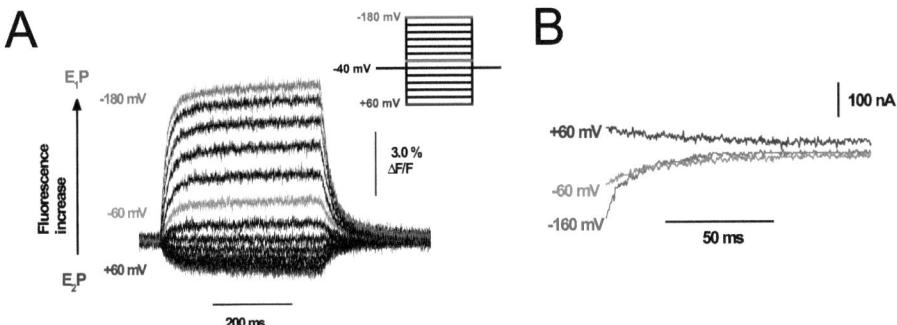

Figure 3.4.: **Voltage-pulse induced fluorescence changes and transient currents of TMRM-labeled Na,K-ATPase.** Fluorescence signals (A) and ouabain-sensitive transient currents (B) of the site-specifically labeled Na,K-ATPase in response to voltage pulses from -40 mV to values between +60 mV and -180 mV in 20 mV steps (A, inset) under extracellularly high Na^+/K^+-free conditions are shown. Data originated from an oocyte coexpressing NaKαwt + NaKβS62C. Traces for voltage jumps to -160 mV, -60 mV and +60 mV are highlighted.

To address the same question for the even more heavily glycosylated gastric H,K-ATPase, wild-type and glycosylation-deficient β-isoforms were coexpressed together with the aforementioned S806C variant of the H,K-ATPase α-subunit. Application of the same voltage steps as in the protocol used for the Na,K-ATPase (Figure 3.4, inset) to TMRM-labeled oocytes expressing these constructs under K^+-free conditions resulted in fluorescence changes as presented in Figure 3.6 A. Provided that Na,K- and H,K-ATPase reaction mechanisms during voltage pulse experiments in the absence of extracellular K^+ are similar, one would also expect a voltage dependence of fluorescence amplitudes ((1-$\Delta F/F$)-V curve) according to a Boltzmann function, implying that the ATPases can be shifted saturatingly between E_1P- and E_2P-states. Thus, the almost linear distribution observed between the -200 mV and +60 mV argues for a reduced voltage sensitivity of the gastric H,K-ATPase compared to the Na,K-ATPase. Although fits of a Boltzmann function to (1-$\Delta F/F$)-V curves are less well defined, since saturation of fluorescence amplitudes cannot be reached within the experimentally accessible voltage range, consistent fit results were obtained on many different cells. The resulting value for the equivalent charge z_q of 0.3 for the gastric H,K-ATPase can be considered as an upper limit, since fits of a Boltzmann function would yield even lower z_q values, if the approximately linear progression of the Q-V curve would extend over an even larger voltage range. As shown in Figure 3.6 B, there are no discernible differences between the voltage dependence of fluorescence amplitudes for the wildtype and the glycosylation-deficient H,K-ATPase mutant when coexpressed with HKαS806C, thus resulting in distributions with similar Boltzmann parameters (see Table 3.2). Likewise, the reciprocal time constants from monoexponential fitting of the fluorescence changes (Figure 3.6 C) are not significantly different for the glycosylation-deficient and glycosylated H,K-ATPases. Notably, however, they are by a factor of about 5 smaller compared to those of the Na,K-ATPase (Figure 3.5 D) and their voltage dependence is also substantially weaker. This fits well with the aforementioned reduced voltage dependence of (1-$\Delta F/F$)-V curves, reflected by the much lower z_q value (~0.30 versus ~0.75 for the Na,K-ATPase, see Table 3.2). Together, these two phenomena may explain why all our attempts failed to measure transient currents on H,K-ATPase-expressing oocytes using the two-electrode voltage-clamp or the giant excised patch-clamp technique (unpublished observations). Since the electrogenicity of both the H^+-transporting (van der Hijden et al., 1990) and the K^+-transporting limb (Lorentzon et al., 1988) of the H,K-ATPase catalytic cycle has been demonstrated, such currents probably exist, but the at least 2 to 3 times lower amount of charge transported with approximately 5-fold lower reciprocal time

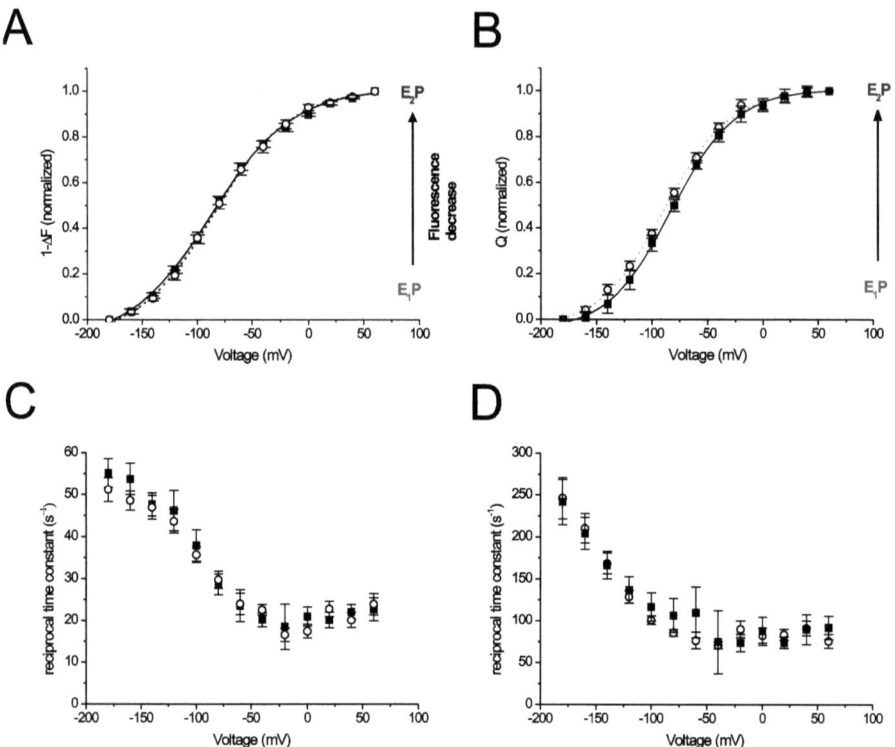

Figure 3.5.: **Voltage dependence of E_1P/E_2P distribution and kinetics of transitions for glycosylated or nonglycosylated Na,K-ATPase enzymes.** Voltage dependence of fluorescence amplitudes 1-$\Delta F/F$ (A) and transient charge movements Q (B) for Na,K-ATPase complexes containing glycosylated (filled squares) or glycosylation-deficient (open circles) β-subunits. Data are means ± S.E. of 10-24 oocytes, normalized to saturating values at -180 mV after subtracting the values for +60 mV. Reciprocal time constants of voltage jump-induced fluorescence changes (C) and transient currents (D) for Na,K-ATPase complexes containing glycosylated or non-glycosylated β-subunits (conditions as in A). Data are means ± S.E. of 8-15 oocytes.

constants would result in substantially smaller current amplitudes compared to the sodium pump, which are difficult to resolve in voltage-jump experiments.

3.3. Discussion

In this chapter, the role of N-linked carbohydrates in the function of Na,K- and H,K-ATPase was examined in *Xenopus* oocytes. The results were surprisingly similar for both investigated enzymes, with no discernible differences between the glycosylation-deficient variants and the fully glycosylated enzymes. Plasma membrane delivery, association with the α-subunit and ion transport activities were virtually unaffected by the lack of N-linked glycans. Moreover, apparent ion affinities were unaltered and even intricate details of the catalytic cycle, such as the voltage-dependent distribution between E_1P and E_2P-states, as well as the kinetics of the voltage-dependent $E_1P \rightarrow E_2P$ transition

3.3. Discussion

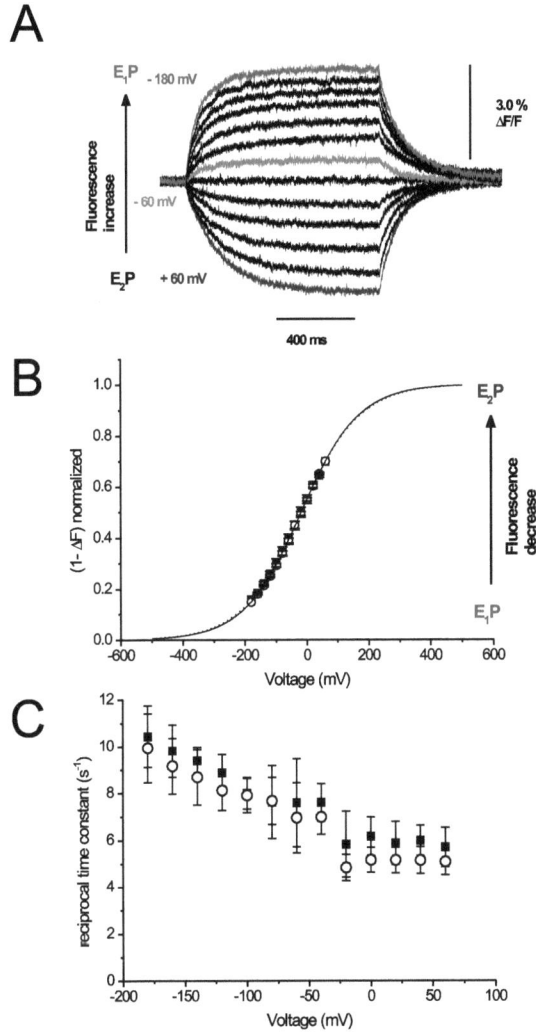

Figure 3.6.: **E_1P/E_2P distribution and kinetics of the $E_1P \rightarrow E_2P$ transition for wildtype or nonglycosylated H,K-ATPase complexes at pH 7.4.** A, Voltage pulse-induced fluorescence responses of the site-specifically labeled H,K-ATPase under K^+-free conditions (pH 7.4). Recordings originated from an oocyte coexpressing HKαS806C + HKβwt. B, Voltage dependence of fluorescence amplitudes (1-ΔF/F) under K^+-free conditions for H,K-ATPase construct HKαS806C containing glycosylated HKβwt (filled squares) or nonglycosylated HKβgd (open squares) β-subunits. Data are means ± S.E. of 10-13 oocytes. A curve resulting from a fit of a Boltzmann function is superimposed. The fluorescence amplitudes 1-ΔF/F were normalized to saturation values from the fits. C, Reciprocal time constants of voltage jump-induced fluorescence changes under K^+-free conditions for the H,K-ATPase αS806C construct containing glycosylated HKβwt (filled squares) or nonglycosylated HKβgd (open squares) β-subunits. Data are means ± S.D. from 9-11 oocytes.

construct	(1-ΔF/F)-V	curves	Q-V	curves
	$V_{0.5}$	z_q	$V_{0.5}$	z_q
Na,Kα₁WT/β₁S62C	-86.9± 2.9	0.71 ± 0.02	-88.7± 1.5	0.87 ± 0.04
Na,Kα₁WT/β₁S62Cgd	-84.4± 2.4	0.75 ± 0.03	-82.6± 0.9	0.89 ± 0.03
H,KαS806C/βwt	-19.6± 5.4	0.26 ± 0.02	-	-
H,KαS806C/βgd	-18.9 ± 4.6	0.27 ± 0.02	-	-

Table 3.2.: **Parameters from fits of Boltzmann functions to Q-V distributions of Na,K-ATPase and to (1-ΔF/F)-V distributions of Na,K-ATPase or H,K-ATPase enzymes with wildtype or glycosylation-deficient β-subunits.** Values are means ± S.E. of 10-24 oocytes from 2-3 oocyte batches.

were essentially the same for glycosylation-deficient Na,K- and H,K-ATPase β-variants when compared to their glycosylated counterparts. While this is in agreement with previous glycosylation studies on Na,K ATPase (Takeda et al., 1988; Tamkun and Fambrough, 1986; Sun and Ball, 1994; Beggah et al., 1997; Zamofing et al., 1989), similar investigations on the gastric H,K-ATPase using other expression systems had resulted in very dissimilar observations:

Whereas α/β coassembly tested by immunoprecipitation was unaffected for fully deglycosylated H,K-ATPase β-subunits in tunicamycin treated Sf9 cells (Klaassen et al., 1997) and single-site glycosylation-deficient variants expessed in HEK293 cells, the amount of coprecipitated α-subunit was substantially lower when more than four glycosylation sites were deleted and was successively reduced with each deleted site in the mammalian expression system (Asano et al., 2000). For the Na,K-ATPase expressed in *Xenopus* oocytes, slightly lower amounts of α-subunit were coprecipitated with a completely glycosylation-deficient mutant than with the glycosylated wildtype β-subunit (Beggah et al., 1997), also indicating a destabilization of the α/β interaction. Yet, the findings from the present study do not hint at a destabilizing effect on the α/β coassembly caused by the lack of carbohydrates for both Na,K- and H,K-ATPase. Although we did not use immunoprecipitation to directly probe the interaction of the glycosylation-deficient β-subunits with their corresponding α-subunits, we drew our conclusions (i) from quantifying the amount of coexpressed α-subunit delivered to the plasma membrane as judged by immunoblotting of the plasma membrane protein fraction and (ii) by measuring the ion transport activity of ATPase molecules present in the plasma membrane of intact oocytes. We think this is an equally legitimate approach, since according to all previously published data, α/β coassembly is an essential prerequisite for plasma membrane delivery of the α-subunit and enzymatic activity of Na,K-ATPase (McDonough et al., 1990; Ackermann and Geering, 1990)) and H,K-ATPase (Gottardi and Caplan, 1993) holoenzymes. Moreover, our results are in agreement with yeast two-hybrid studies which revealed an interaction between the ectodomain of the β-subunit and the extracellular TM7/TM8 loop of the α-subunit (Melle-Milovanovic et al., 1998). Since there is little or no glycosylation of nuclear transcription factors, the yeast two-hybrid system detects interactions in the probable absence of glycosylation. Accordingly, glycosylation of the β-subunit cannot be required for the association with the α-subunit.

In addition to α/β association, also the plasma membrane delivery of the H,K-ATPase was impaired for glycosylation-deficient mutants in previous studies using confocal microscopy or immunofluorescence. Whereas single or double glycosylation site-deleted H,K-ATPase β-subunits still supported some α-subunit cell surface expression, deletion of all glycosylation sites prevented plasma membrane delivery in COS-1 cells (Asano et al., 2000). In HEK293 cells however, every single-site deletion already abolished plasma membrane delivery, except for the deletion of the nonconserved glycosylation site Asn103 (Vagin et al., 2003). In the Sf9 expression system, the situation is even more complicated due to the fact that even the normally glycosylated, catalytically active wildtype holoenzyme was not found in the plasma membrane of insect cells, but had to be retrieved from intracellular membrane structures. However, the fully glycosylated wildtype β-subunit was partly targeted to the plasma membrane, even when the α-subunit was not coexpressed, whereas plasma membrane delivery was not observed after a complete removal of N-linked sugars by tunicamycin treatment of Sf9 cells (Klaassen et al., 1997). Therefore, the authors concluded that plasma membrane delivery of the α-subunit depends on N-linked glycosylation. Eventually, ATPase activity was completely lost both for tunicamycin-treated, H,K-ATPase-transfected insect cells (Klaassen et al., 1997) and for a fully N-glycosylation-deficient H,K-ATPase variant expressed in HEK293 cells according to Asano and co-workers (Asano et al., 2000). Single or double site-deleted mutants still exhibited normal activity in HEK293 cells, but each additional deletion resulted in decreased activity (Asano et al., 2000). Similarly, Vagin and co-workers reported unaltered enzymatic properties and SCH 28080 inhibition kinetics for every single- and some double-deletion variant stably expressed in HEK cells (Vagin et al., 2003).

These observations disagree with our results that rubidium transport activity of the glycosylation-deficient variant is not different from glycosylated H,K-ATPase in *Xenopus* oocytes. In our opinion, this expression system has several advantages: Stationary currents of the Na,K-ATPase and rubidium uptake of the H,K-ATPase were measured on intact plasma membranes from living oocytes, whereas in the other expression systems enzymatic activity was determined in crude membrane fractions, which do not differentiate between plasma membrane and intracellular membranes. For the latter studies it is therefore more likely that the loss of activity and potentially also the impaired α/β coassembly observed for completely deglycosylated H,K-ATPase occurred as a consequence of preparational procedures and would possibly not be observed when determined on intact cells or on pure plasma membrane protein fractions. The lack of sugar moieties seems to render the enzyme more susceptible to inactivation than the glycosylated wildtype isoform during preparational procedures, which apparently does not happen as long as the enzyme is stabilized in a native plasma membrane environment. Results from purification studies on Na,K-ATPase isolated from *Pichia pastoris* support this idea, since it was reported that enzymatic removal of all N-linked carbohydrates from isolated Na,K-ATPase molecules reduced their activity, especially when facing certain ionic conditions and lipid compositions. The presence of some particular phospholipids on the other hand conserved enzymatic activity of the deglycosylated enzyme (Cohen et al., 2005). In this context it is interesting to note that activity measurements in most studies on glycosylation-deficient Na,K-ATPase mutants were also carried out in living cells, mainly in oocytes (Takeda et al., 1988; Beggah et al., 1997). H,K-ATPase activity might even be more sensitive to preparational impairment than that of Na,K-ATPase, considering its high sensitivity to inactivation by detergents (Klaassen et al., 1997). An even more likely explanation for the discrepancies found between glycosylation-deficient H,K-ATPase variants expressed in oocytes and the other two mentioned expression systems might be that the observed differences are temperature-related. It was shown that glycosylation increases the thermal stability of proteins, resulting from largely entropic rather than enthalpic contributions (DeKoster and Robertson, 1997). This might also explain why the α/β interaction of a glycosylation-deficient variant is only disrupted in mammalian cells cultured at 37 °C, but not in oocytes or Sf9 cells, which are incubated at 17 and 27 °C, respectively. Minor folding impairments potentially occurring already at 27 °C might not affect α/β coassembly but still abolish enzymatic activity, possibly explaining the loss of H,K-ATPase activity reported for the Sf9 expression system.

Although the oocyte system may be regarded as physiologically less relevant, the observation of unimpaired H,K-ATPase activity of glycosylation-deficient mutants in the current study is valuable, because the role of N-linked sugars on enzyme function can be clarified. Our findings of a completely unaffected surface delivery of glycosylation-deleted mutants are rather incompatible with the idea that the carbohydrates are essential for correct folding and intersubunit association for both Na,K- and H,K-ATPase. However, they are not contradicting the role of the N-glycans for basolateral sorting reported for Na,K-ATPase complexes containing $β_1$-subunits (Lian et al., 2006) versus apical sorting for Na,K-ATPase holoenzymes comprising $β_2$-subunits (Vagin et al., 2005a), and H,K-ATPase with fully glycosylated β-subunits (Vagin et al., 2004). In the nonpolarized oocyte, such targeting signals may not be recognized. For instance, a Y20A mutation in the basolateral sorting/endocytosis signal sequence FRXY present in the H,K-ATPase β-subunit was shown to enhance apical targeting of the H,K-ATPase *in vivo* (Courtois-Coutry et al., 1997; Wang et al., 1998) and in polarized MDCK, but not LLC-PK cells (Roush et al., 1998). In contrast, immunoblotting of isolated plasma membranes or Rb^+ uptake measurements did not reveal any differences in the amount of active H,K-ATPase in the plasma membrane, when *Xenopus* oocytes expressing the α-subunit together with the same β-variant Y20A or the wildtype β-subunit were compared (data not shown). Furthermore, removal of the tyrosine-based targeting signal by an extensive N-terminal truncation of the H,K-ATPase β-subunit (HKβΔ29 mutant in chapter 5) did neither impair plasma membrane targeting nor α/β coasssembly (see Figure 5.2 on page 70).

Apart from their importance as apical sorting signals in polarized cells, the N-linked carbohydrates of the gastric H,K-ATPase β-subunits may have a protective role *in situ*. It was shown that the composition of terminal sugars of the H,K-ATPase oligosaccharides in parietal cells is quite different from other glycosylated proteins (e.g., absence of sialic acid (Tyagarajan et al., 1996; Beesley and Forte, 1973) and possibly designed to resist hydrolysis of the carbohydrate chains in the acidic stomach lumen. This in turn raises the question, why the integrity of the carbohydrates is important for the gastric H,K-ATPase, once delivered to the plasma membrane. Interestingly, it was shown that the removal of N- or O-linked carbohydrates destabilizes a variety of proteins and tends to make them more susceptible to aggregation, especially at low pH (Wang et al., 1996; Imperiali and O'Connor, 1999). Moreover, results from several studies indicate that the removal of oligosaccharides renders the H,K-ATPase more sensitive toward proteolytic digestion by various enzymes including pepsin (Thangarajah et al., 2002; Crothers et al., 2004). Notably, a similar protective role of glycosylation has been demonstrated for other gastrointestinally active enzymes (Rudd et al., 1994; Loomes et al., 1999).

In summary, the results from this study on *Xenopus* oocytes favor the idea that in analogy to the Na,K-ATPase, glycosylation of the H,K-ATPase is dispensable for protein stability and enzyme function itself, but is rather important for other functions primarily relevant in native tissues.

CHAPTER 4

Functional Significance of E_2-specific transmembrane interactions between α- and β-subunits of Na,K- and H,K-ATPase

Dürr et al. (2009) The Journal of Biological Chemistry 284, 3842–3854

4.1. Introduction

Apart from their important chaperone-like functions including folding, membrane insertion and plasma membrane delivery of the catalytically active α-subunits (Gottardi and Caplan, 1993; McDonough et al., 1990; Geering, 1991)), the β-subunits of sodium and proton pumps were shown to be essential for enzyme activity (Noguchi et al., 1987)) and to influence the transport properties of the mature enzymes. The existence of various tissue-specific Na,K-ATPase β-subunit isoforms, which result in holoenzymes with different cation affinities (Jaisser et al., 1994; Crambert et al., 2000), and the fact that Na,K-ATPase α-subunits coexpressed with H,K-ATPase β-subunits form active pumps (Horisberger et al., 1991) albeit with altered ion affinities (Jaisser et al., 1994; Eakle et al., 1992), strongly suggests a modulatory function of the β-subunit for ion translocation.

Results from several studies indicate that it is mainly the C-terminal extracellular domain of the β-subunit which modulates cation transport. For example, reduction of the disulfide bonds in the β-subunit's ectodomain impairs the function of purified Na,K- or H,K-ATPase. Since this was prevented in the presence of cations (Kawamura et al., 1985; Lutsenko and Kaplan, 1993; Chow et al., 1992), a potential role of the β-subunit in K^+ occlusion has been suggested. Functional analysis of chimeras between Na,K- and H,K-ATPase β-subunits confirmed that mostly α/β ectodomain interactions are responsible for the observed effects of the β-subunit on cation affinity and occlusion of the sodium pump (Jaunin et al., 1993; Eakle et al., 1994; Hasler et al., 1998). This is also in line with the electron density of the first 10-15 residues of the ectodomain, which was tentatively traced in the X-ray structure of the pig Na,K-ATPase holoenzyme (Morth et al., 2007) and clearly indicates that the extracellular M5/M6 and M7/M8 loops of the α-subunit are covered by a "lid" formed by the β-subunit's ectodomain (see Figure 1.3 on page 11), as suggested previously (Lutsenko and Kaplan, 1993). Molecular details of these extracellular α/β interactions were recently revealed by the shark rectal gland Na,K-ATPase structure at 2.4 Å resolution (Shinoda et al., 2009).

Yet, not only ectodomain modifications have been shown to alter the transport properties. N-terminal truncation of the β-subunit's cytoplasmatic domain of the Na,K-ATPase resulted in changes of the apparent K^+ (Hasler et al., 1998; Geering et al., 1996) and Na^+ affinities (Hasler et al., 1998) and affected the conformational equilibrium (Abriel et al., 1999). Likewise, an inhibitory antibody, which recognizes an epitope within the first 36 N-terminal amino acids of the H,K-ATPase's β-subunit, altered the K^+affinity (Chow and Forte, 1993). However, cytoplasmic interactions are probably not directly responsible for the functional effects of β-subunits on cation binding of the sodium pump, since in contrast to a complete truncation, deletions or multiple mutational alterations of the N-terminus did not affect the K^+ activation of Na,K-ATPase expressed in *Xenopus* oocytes (Hasler et al., 1998). Furthermore, results from a glycosylation mapping assay indicated a repositioning of the transmembrane segment as a consequence of the N-terminal truncation (Hasler et al., 2000). This in turn may also impair the conformation of the ectodomain, whose significance for cation occlusion has already been outlined above. Even the observed repositioning of the transmembrane domain itself could be responsible for the reported K^+ effects of the N-terminally truncated β-variant, since mutations in the TM have also been shown to modify cation transport: tryptophan scanning mutagenesis in the Na,K-ATPase β-subunit's TM revealed that the replacement of two tyrosines by tryptophan has distinct and additive consequences for the holoenzyme's cation affinities (Hasler et al., 2001). Interestingly, these tyrosines are highly conserved, actually being present in all known β-subunits (represented by red capital letters in the β-TM alignment shown in Figure 4.1). The apparent $K_{0.5}$ for extracelluar K^+ activation of pump currents in *Xenopus* oocytes was significantly increased for a simultaneous tryptophan replacement of the two tyrosines in various β-subunit isoforms. Of note, this was accompanied by an increase of the apparent affinity for intracellular sodium and a reduced sensitivity towards the E_2-specific inhibitor vanadate. Therefore, it was concluded that the affinity changes might occur secondary to a conformational, i.e. E_2-destabilizing effect of the β-subunit mutations (Hasler et al., 2001). However, direct evidence for a shift in the enzyme's E_1P/E_2P distribution has not been provided yet. Given the high propensity of native gastric H,K-ATPase to occur in the E_2-state (Helmich-de Jong et al., 1985), which has been speculated to be of primary importance for efficient H^+ delivery to the luminal fluid against a 10^6-fold H^+ gradient *in vivo*, it is of high interest to identify possible interaction

4.1. Introduction

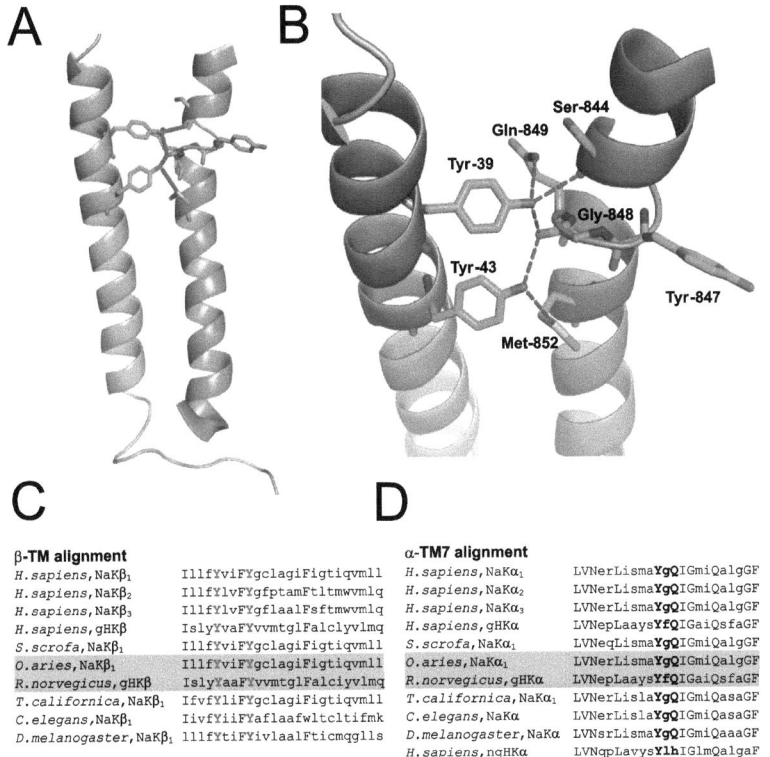

Figure 4.1.: **Structural representation of the transmembrane α/β interface of pig renal Na,K-ATPase and alignments of several Na,K- and H,K-ATPase α-TM7 and β-TM.** A, B: Illustration of possible interaction sites between β-TM and α-TM7 of Na,K-ATPase, according to the 3B8E crystal structure (Morth et al., 2007). Putative hydrogen bonds are shown to a cut-off-value of 3.8 Å. B: A close-up view of the β-TM/α-TM7 interface, showing the two conserved tyrosines Y39 and Y43 of the β-subunit and some selected α-subunit residues located at interacting distance. C, D: alignments of β-TM(C) and α-TM7(D) of Na,K- and H,K-ATPases from different species or of different human isoforms. Highly conserved[1] amino acids are represented by capital letters; residues altered by mutagenesis in the current study are shown in bold letters.

sites on the β- and ultimately also on the α- subunit's transmembrane domains, which are crucial for this unique E_2-specific structural stabilization.

---

[1] present in at least 75% of the investigated sequences according to manually refined alignments performed by Axelsen and Palmgren (1998). See supplementary data (Chapter A, page 86-89) for a full β-subunit alignment.

In this chapter we set out to identify molecular determinants of E_2-specific inter-subunit interactions and utilized the technique of voltage-clamp fluorometry to directly determine the distribution between E_1P- and E_2P-states of wildtype and mutated Na,K- and H,K-ATPase enzymes (Geibel et al., 2003b; Dempski et al., 2005). Since this method can be applied to the electrogenic Na,K-ATPase as well as to the electroneutrally-operating gastric H,K-ATPase (Geibel et al., 2003a), it can be investigated whether mutation of the conserved tyrosines causes similar conformational shifts in both enzymes. Evidence is provided that the observed conformational effects are of a more general significance for ion translocation by oligomeric P-type ATPases. Furthermore, we compared for both enzymes the effects of these β-subunit mutations on apparent cation affinities, which occur in conjunction with shifts in the distribution of E_1P/E_2P conformational states. Moreover, we studied the effect of mutations of selected residues in the TM7 of the Na,K- and H,K-ATPase α-subunit, which are at interacting distance to the two β-subunit tyrosines (Figure 4.1). This strategy provided novel insights about the molecular details of interactions between β- and α-subunit transmembrane domains of oligomeric P_{2C}-type ATPases, which are responsible for stabilization of the E_2 conformational state.

4.2. Results & Discussion

4.2.1. E_1P/E_2P conformational distribution and kinetics of the E_1P/E_2P-transition for Na,K-ATPase wildtype and β—(Y39W,Y43W) mutant enzymes

To determine whether the double mutation Y39W/Y43W in the Na,K-ATPase β_1-subunit causes the proposed shift of the enzyme's conformational equilibrium towards E_1P (Hasler et al., 2001), the two tryptophans were introduced into the reporter construct β_1-S62C and coexpessed with the α_1-subunit in *Xenopus* oocytes. Upon labelling with the environmentally-sensitive dye TMRM, we applied voltage-jumps under extracellularly high Na^+/K^+-free conditions. Under these conditions, the sodium pump carries out Na^+/Na^+ exchange where the enzyme shuttles exclusively between E_1P/E_2P-states of the catalytic cycle. The ratio E_1P/E_2P is increased by extracellular Na^+, but the effect is opposed if extracellular K^+ is also present (Kaplan and Hollis, 1980; Kaplan, 1982). Figure 4.2 shows fluorescence signals and transient currents in response to three representative voltage-steps (-160 mV, -60 mV, +60 mV) for an oocyte expressing the Na,K-ATPase α_1-subunit together with either the reference construct β_1-S62C (4.2 A and C, respectively) or the mutated β_1-subunit variant -β_1S62C(Y39W,Y43W) (4.2 B and D, respectively).

Of note, voltage-jumps to extremely hyperpolarizing potentials (-160 mV), which force the sodium pump into the E_1P-state, result in a relative fluorescence increase, as described in (Dempski et al., 2005), which is more pronounced for the wildtype (Figure 4.2 A) than for the mutant enzyme (Figure 4.2 B). In contrast, E_2P-promoting positive membrane potentials lead to a relative fluorescence decrease, which is substantially larger for the double-Trp mutant than for the non-mutated construct. This indicates that e.g. at -40 mV the dynamic equilibrium between the two principal conformations is shifted towards E_1P for the β-variant enzyme. This shift of the voltage-dependent distribution is also reflected by the concomitantly recorded transient currents of the mutant, which exhibit larger amplitudes for voltage jumps to positive potentials (e.g. +60 mV, Figure 4.2 D) compared to the corresponding wildtype signals (Figure 4.2 C).

Both, the resulting voltage-dependent distribution of transported charge (Q-V curve, Figure 4.3 A) and the (1-ΔF/F)-V distribution (Figure 4.3 B) can be fitted to a Boltzmann function, yielding midpoint potentials which are significantly more positive for the β_1-S62C (Y39W,Y43W) mutant than for the β_1-S62C reference enzyme (see Table 4.1 on page 56). Moreover, the voltage-dependent reciprocal time constants (τ^{-1}) of transient currents (Figure 4.3 C) and of fluorescence changes (Figure 4.3 D) are also shifted by almost +50 mV towards depolarizing potentials for the variant sodium pumps. Notably, the reciprocal time constants are significantly larger for the mutant enzyme at negative potentials but not at positive potentials. This indicates that only the apparent rate constant for reverse binding of extracellular Na^+ and the subsequent conformational change from E_2P to E_1P is increased for the mutant. This is in agreement with the increased apparent affinity for extracellular Na^+ as reported by Hasler et al. (2001), which was most pronounced at hyperpolarizing potentials

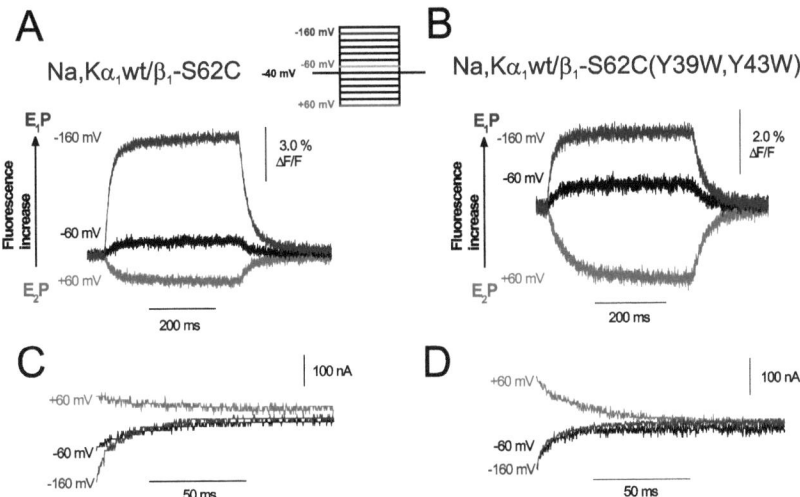

Figure 4.2.: **Voltage-pulse induced fluorescence changes and transient currents of site-specifically labeled Na,K-ATPase wildtype and β-(Y39W,Y43W) mutant enzymes.** Fluorescence amplitudes (A,B) and transient currents (C,D) in response to voltage pulses from -40 mV to three representative voltages (+60 mV, -60 mV, -160 mV, see inset in A) under extracellularly high Na^+/K^+-free conditions are shown. Data originated from individual oocytes coexpressing the wildtype Na,K-ATPase α_1-subunit with either the unmodified reporter construct β_1S62C (A,C) or the reporter construct β_1S62C (Y39W,Y43W) carrying tryptophan replacements of the two conserved tyrosines (B, D).

(as inferred from the sodium-dependent inhibition of pump currents especially at hyperpolarizing membrane potentials). The increased apparent rate constant for the backward reaction sequence (extracellular Na^+ reverse binding / $E_1P \rightarrow E_2P$ transition) observed for the mutant in the current study can also account for the mutant's increased apparent $K_{0.5}$ for extracellular K^+, which was shown to be more pronounced at high extracellular Na^+ concentrations (Hasler et al., 2001). With increasing extracellular K^+ concentrations a relative decrease of the fluorescence amplitudes $\Delta F/F$ of TMRM-labeled Na,K-ATPase is observed (Figure 4.4 A-E). Thus, the apparent $K_{0.5}$ for extracellular K^+ can be determined from the K^+-dependence of this decrease, as already demonstrated by Dempski et al. (2005). Using this independent approach, we confirm here the reduced extracellular K^+ affinity for the mutant in presence of Na^+ (Figure 4.4 F).

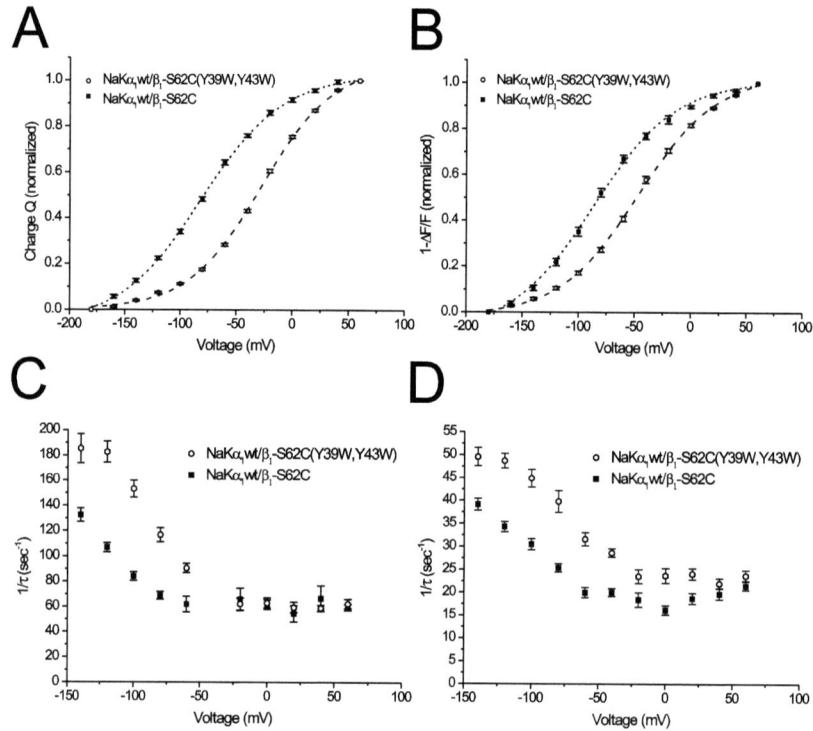

Figure 4.3.: **Voltage dependence of the E_1P/E_2P distribution and kinetics of the E_1P/E_2P transition for Na,K-ATPase wildtype and (Y39W,Y43W) β-variant enzymes.** A, B: Voltage-dependent distributions of transported charge Q (A) and of fluorescence amplitudes 1-$\Delta F/F$ (B) for Na,K-ATPase consisting of the wildtype α_1-subunit and either unmodified reporter construct β_1-S62C, or mutated β_1-S62C(Y39W,Y43W). Data are means ± S.E. of 20-25 oocytes from 3-4 different oocyte batches, normalized to saturating values at -180 mV after subtracting the values for +60 mV. A curve corresponding to the fit of a Boltzmann function is superimposed. C, D: Reciprocal time constants of voltage-jump induced transient currents (C) and fluorescence changes (D) for oocytes expressing the Na,K-ATPase wildtype α_1-subunit together with either β_1S62C or β_1S62C(Y39W,Y43W). Data are means ± S.E. of 10-15 (C) or 20-25 (D) oocytes from 3-4 different oocyte batches respectively.

4.2.2. Molecular determinants for the E_2-stabilizing effect mediated by the two conserved tyrosines of the β-subunit

To further clarify which particular property of the β−TMD tyrosines is necessary to maintain the observed E_2P-preference in the E_1P/E_2P distribution within the reference construct β_1−S62C, we determined the E_1P/E_2P conformational distribution of additional β_1-S62C variants: β_1−S62C(Y39S,Y43S) was chosen to conserve the hydroxyl group, and β_1−S62C(Y39F,Y43F) to maintain the phenyl moiety of the side chains. According to the Boltzmann parameters obtained for the (1-$\Delta F/F$)-V curves of these two mutants (Table 4.1 on page 56), a significant shift of the conformational equilibrium towards the E_1P-state is observed for both, but the effect is more pronounced for the serine replacements. This

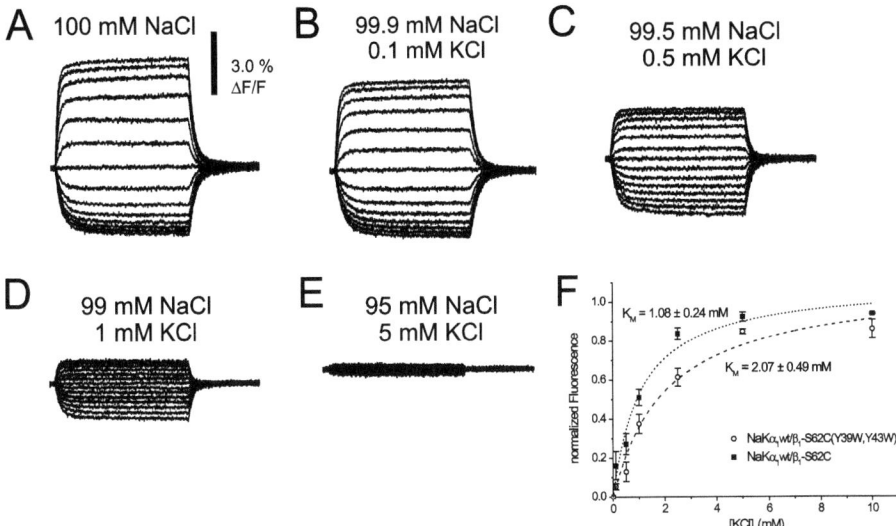

Figure 4.4.: **Determination of the apparent $K_{0.5}$ for extracellular K^+ of Na,K-ATPase $\alpha_1 wt/\beta_1 S62C$ and $\alpha_1 wt/\beta_1 S62C(Y39W,Y43W)$ variant enzymes by fluorescence titration experiments.** A-E: Representative fluorescence titration experiment of a single oocyte expressing $\alpha_1 wt/\beta_1 S62C$. Voltage-jumps were applied from a holding potential of -40 mV to potentials between +60 mV and -180 mV at different external K^+ concentrations (as indicated) in presence of Na^+. F: K^+ activation curves resulting from titration experiments on $\alpha_1 wt/\beta_1 S62C$ and $\alpha_1 wt/\beta_1 S62C(Y39W,Y43W)$ Na,K-ATPase expressing oocytes. Data are means ± S.E. of 3-5 independent experiments.

indicates that both the hydroxyl group and the aromatic phenyl moiety may be involved in the interaction, but surprisingly the aromatic ring appears to be more important. Apparently, not only "classical" H-bonds (illustrated in Figure 4.1 A and B) are responsible for the E_2-stabilizing effect of the two tyrosines, but possibly also amino-aromatic interactions (or amino-aromatic H-bonds) with side-chain or even backbone amides in the α–TM7 (see below). Notably, according to Hasler et al. (2001), a significantly reduced apparent K^+ affinity was only observed for β-(Y39S,Y43S) but not for β-(Y39F,Y43F), which corroborates this interpretation. Our results also demonstrate that the fluorometric technique applied here is a substantially more sensitive method. It is able to detect even subtle changes in the E_1P/E_2P poise, which may have no direct consequences on the apparent ion affinities, but still are relevant to identify the molecular determinants for the interaction patterns of the tyrosine side chains.

4.2.3. E_1P/E_2P-conformational distribution, apparent Rb^+ affinities and SCH 28080 sensitivity of the H,K-ATPase wildtype and β-(Y44W,Y48W) variant enzymes

The study by Hasler et al. (2001) also included a chimeric β-subunit consisting of the cytoplasmic and transmembrane domains of the Na,K-ATPase β_1-subunit and the ectodomain of the gastric H,K-ATPase β-subunit, which was co-expressed with the Na,K-ATPase α-subunit. Since the tryptophan mutations were accordingly introduced into a TM derived from the Na,K-ATPase β_1-subunit and its effects were monitored by Na,K-ATPase transport activity, this chimeric ATPase does not represent an optimal system to clarify the functional relevance of the conserved tyrosines for H,K-ATPase ion transport activity. We therefore extended the current study to the true oligomeric form of the gastric H,K-ATPase by co-expressing the H,K-ATPase β-subunit with an α-subunit variant carrying a

	Boltzmann parameter $(1-\Delta F/F)$-V curves		Boltzmann parameter Q-V curves	
construct	$V_{0.5}$	z_q	$V_{0.5}$	z_q
Na,K$\alpha_1$wt/β_1S62C	-94.8 ± 4.8	0.74 ± 0.01	-76.6 ± 4.8	0.77 ± 0.03
Na,K$\alpha_1$wt/β_1S62C(Y39W,Y43W)	-44.8 ± 1.5	0.73 ± 0.03	-29.3 ± 0.9	0.79 ± 0.03
Na,K$\alpha_1$wt/β_1S62C(Y39F,Y43F9)	-79.1 ± 3.6	0.72 ± 0.02	n.d.	n.d.
Na,K$\alpha_1$wt/β_1S62C(Y39S,Y43S)	-56.1 ± 8.0	0.74 ± 0.03	n.d.	n.d.
H,KαS806C/βwt*	-125.1 ± 11.4	0.49 ± 0.07	-	-
H,KαS806C/β-(Y44W,Y48W)*	-90.0 ± 3.4	0.68 ± 0.06	-	-

* determined in 90 mM TMACl pH 5.5

Table 4.1.: **Parameters from fits of a Boltzmann function to Q-V distributions of Na,K-ATPase and $(1-\Delta F/F)$-V distributions of Na,K-ATPase or H,K-ATPase β-TM-variants.** Values are means ± S.E. of 15-25 oocytes from 3-4 oocyte batches.

single cysteine replacement (S806C) in the M5/M6 loop, which is necessary to enable voltage-clamp fluorometry (Geibel et al., 2003a) but does not impair enzyme activity (see section 2.1.1, Figure 2.1 on page 32).

Figure 4.5 A shows the voltage dependence of fluorescence changes obtained for oocytes co-expressing the H,K-ATPase α-S806C variant together with either the wildtype H,K β-subunit or a H,K-ATPase β-subunit carrying the (Y44W,Y48W) mutation. In analogy to Na,K-ATPase, the voltage dependence of fluorescence amplitudes observed for the H,K-ATPase β-(Y44W,Y48W) is shifted towards depolarizing potentials compared to proteins comprising wildtype β-subunits. The resultant Boltzmann distribution is characterized by a substantially smaller slope z_q than the corresponding Na,K-ATPase curves (see Table 4.1), thus indicating a reduced voltage sensitivity, as already described in the previous chapter. This can be interpreted as a reduced dielectric depth of a so-called 'ion access channel' which protons (or hydronium ions) pass upon voltage pulses in order to reach or exit from the cation binding sites. Since a smaller fraction of the transmembrane potential is accordingly sensed by the transported charge, the process is less electrogenic than the extracellular Na$^+$ binding/release, which indicates a deeper ion well in case of the sodium pump (Sagar and Rakowski, 1994).

Since full saturation of fluorescence amplitudes cannot be reached in the experimentally accessible voltage range, the saturating values of the Boltzmann curves had to be determined by extrapolation of the fit function. Of note, only small fluorescence changes ($\Delta F/F$~0.5-2%) were consistently observed for the β−(Y44W,Y48W) variant, corresponding to about 10% of the wildtype fluorescence signals ($\Delta F/F$~5-20%). To account for this, the saturation level used for normalization of the mutant has been arbitrarily set to 0.1 (instead of 1 for the wildtype), which is illustrated by the different scaling of the y-axis (right axis in Figure 4.5 A).

To determine whether the conformational shift towards E$_1$P observed here for the H,K-ATPase β-(Y44W,Y48W) mutant also causes changes in cation affinities, we determined the apparent Rb$^+$ affinity in Rb$^+$ uptake experiments. As illustrated in Figure 4.5 B, the apparent affinity of the variant enzyme for extracellular Rb$^+$ in absence of extracellular Na$^+$ is drastically reduced by a factor of more than 20. If the apparent K$_{0.5}$ is determined in the presence of extracellular Na$^+$ (Figure 4.5 C), the apparent K$_{0.5}$ values are substantially larger for both the wildtype and the β-(Y44W,Y48W) variant enzyme, but the affinity decrease caused by the double-Trp mutation is still observed, albeit less pronounced. Interestingly, the apparent K$_{0.5}$ of the wildtype H,K-ATPase is more strongly influenced by the presence of Na$^+$, since it is almost 7-fold higher than in absence of Na$^+$, whereas the apparent K$_{0.5}$ of the mutant is increased by a factor of less than 1.5.

Considering the E$_1$P-shift observed for the mutant enzyme (see above), these findings have a reasonable explanation: The effects of sodium ions on the H,K-ATPase have been described in various studies (Ray and Nandi, 1985; Polvani et al., 1989; Rabon et al., 1990; Swarts et al., 1995) and it was shown that they compete with K$^+$ binding to the extracellular binding sites (Swarts et al., 1995). Therefore the wildtype enzyme, which is predominantly in the E$_2$P-state is more susceptible to this competition of Na$^+$ ions, which in effect reduces the apparent Rb$^+$ affinity in the presence of Na$^+$. In contrast, the

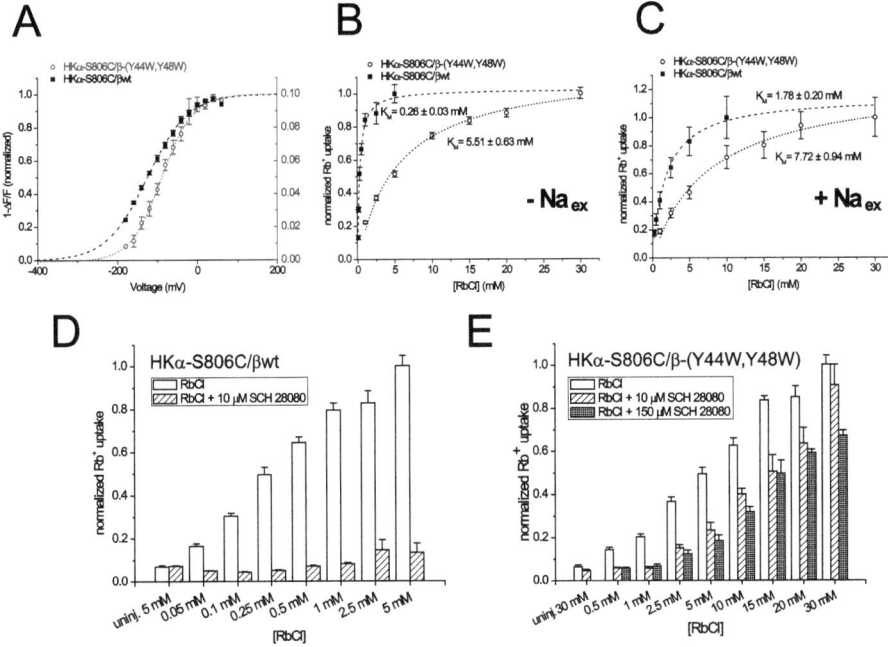

Figure 4.5.: **Functional properties of the gastric H,K-ATPase βY44W,Y48W mutant enzyme.** A: Voltage-dependent E_1P/E_2P distribution of fluorescently labeled αS806C/βwt and αS806C/β(Y44W,Y48W) H,K-ATPase enzymes, resulting from fluorescence responses upon voltage-jumps from a holding potential of -40 mV to values between +60 mV and -180 mV (20 mV increments) at pH 5.5. Data are means ± S.E. of 10-15 oocytes from 2-3 different oocyte batches. A curve resulting from a fit of a Boltzmann function is superimposed. Fluorescence amplitudes 1-$\Delta F/F$ were normalized to saturation values from the fits. To account for the substantial differences in the fluorescence amplitudes, the saturation values were set to 1 for the wildtype and to 0.1 for the Y44W,Y48W mutant respectively. B, C: Michaelis-Menten plots for concentration-dependent Rb^+ uptake by H,K-ATPase in absence (B) or presence (C) of extracellular Na^+ at pH 5.5. Oocytes were injected with αS806C mutant and the wildtype (filled squares) or Y44W,Y48W variant β-construct (open circles), respectively. Data were normalized to Rb^+ uptake at saturating RbCl concentrations, corresponding to values between 20-30 pmol min^{-1}/oocyte for different oocyte batches (αS806C/βwt: 29.3, 27.4, 30.8 in absence of Na^+; 21.1, 23.4 in presence of Na^+; αS806C/β(Y44W,Y48W): 28.4, 33.9 in absence of Na^+; 32.6, 30.1 in presence of Na^+). Data are means ± S.E., n = 8-12 oocytes from two or three independent experiments. Apparent half-maximal activation constants $K_{0.5}$ (in mM) were obtained from a fit of a Michaelis-Menten-type function to the data (dashed lines). D, E: SCH 28080 sensitivity of Rb^+ uptake by oocytes expressing αS806C/βwt (D) or αS806C/β(Y44W,Y48W) (E) H,K-ATPase complexes. Rb^+ uptake was determined at pH 5.5 in extracellular Na^+-free solutions containing different Rb^+ concentrations in absence (white bars) or presence of SCH 28080 (hatched bars: 10 μM, crossed bars: 150 μM). Data are means ± S.E., n = 10-15 oocytes from at least two independent sets of experiment, normalized to Rb^+ uptake at saturating RbCl concentrations.

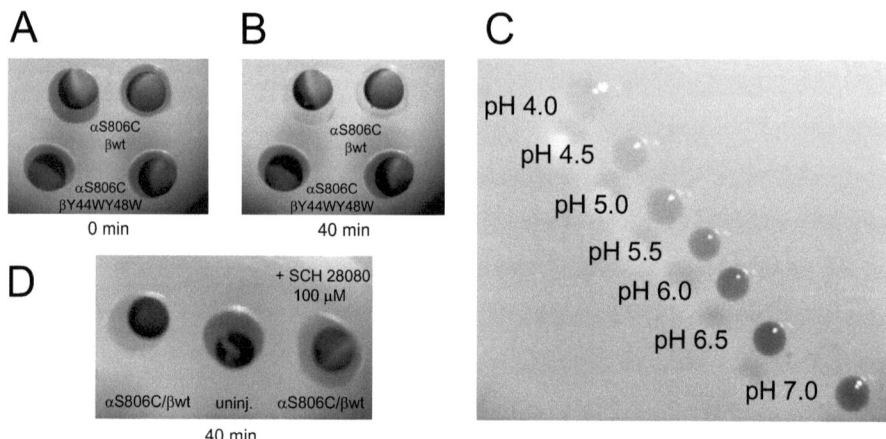

Figure 4.6.: **Acidification assay for gastric H,K-ATPase-expressing oocytes.** A: Oocytes co-expressing the H,K-ATPase αS806C construct with either the wildtype or the (Y44W,Y48W) β-construct were placed under mineral oil in a droplet of weakly buffered solution at pH 7.4 containing pH indicator phenol red. B: After 40 minutes, a substantial acidification to at least pH 4 is observed for oocytes expressing αS806C/βwt H,K-ATPases (B, upper two oocytes), but not for those expressing the β-variant (B, lower two oocytes). One representative out of several similar experiments is shown. C: Extracellular solutions adjusted to different pH containing phenol red to provide a colorimetric scale. D: Control experiments on uninjected or HKαS806C/βwt expressing oocytes in presence of 100 μM SCH 28080 after 40 min incubation.

β-(Y44W,Y48W) variant enzyme is shifted to the E_1P-state and thus cannot efficiently bind extracellular Rb^+. Accordingly, this mutant exhibits a substantially reduced apparent Rb^+ affinity, irrespective of the presence of Na^+ ions, which exert their competitive effect only in the E_2P-state. Figure 4.5 E illustrates another interesting property of the H,K-ATPase β-(Y44W,Y48W) mutant: It exhibits a strongly reduced sensitivity towards the inhibitor SCH 28080. Whereas for the wildtype enzyme Rb^+ uptake is readily inhibited by the compound already at concentrations as low as 10 μM (to about 10-15 % residual activity at saturating 5 mM RbCl, see Figure 4.5 D), the variant is only partially inhibited to residual activities of 30-80% depending on the extracellular Rb^+ concentration, even at higher SCH 28080 concentrations up to 150 μM (Figure 4.5 E). Since SCH 28080 is known to be an E_2-specific inhibitor (see panel B in Figure 1.10 on page 27 or Wallmark et al., 1987), the observed insensitivity of the variant enzyme again reflects its preference for the E_1P-state, which is in close analogy to the reduced vanadate sensitivity of the homologous Na,K-ATPase mutation β-(Y39W,Y43W) described by Hasler and coworkers (Hasler et al., 2001). Control experiments showed, however, that Rb^+ uptake by the variant H,K-ATPase is inhibited to a similar extent as uptake of the wildtype enzyme by omeprazole (data not shown). Since this inhibitor binds covalently to the proton pump (Lorentzon et al., 1985), steady-state inhibition caused by the compound is not conformation-specific.

4.2.4. H^+ secretion of H,K-ATPase wildtype and β-(Y44W,Y48W) variant enzymes

The data presented here suggest a common E_2P-stabilizing effect of the two conserved β-subunit tyrosines in both Na,K- and H,K-ATPase, which is apparently disrupted by the tryptophan replacement in the β-variants. Therefore, as argued for the accelerated apparent rate constant for reverse binding of extracellular Na^+ and the subsequent E_2P/E_1P conformational shift for the $β_1$-(Y39W,Y43W)

Na,K-ATPase variant, in case of the H,K-ATPase containing a double-Trp-mutated β-subunit the observed E_1-shift should have a similar effect on extracellular H^+ reverse binding / E_2P/E_1P conversion. This in fact would kinetically interfere with luminal H^+ release from the external binding sites. To test this hypothesis, we performed a simple assay to evaluate extracellular acidification of wildtype or mutant H,K-ATPase expressing oocytes. Whereas in wildtype H,K-ATPase-expressing oocytes, the extracellular medium is substantially acidified (to at least pH 4) within 40 minutes, as indicated by the distinct change in phenol red color (upper two oocytes in Figure 4.6 A and B), no pH change is observed for oocytes expressing the β-(Y44W,Y48W) mutation (lower two oocytes in Figure 4.6 A and B), even when monitored for longer time periods. Please note that the concentration of RbCl in the extracellular medium was as high as 20 mM, thus representing saturating conditions even for the Y44W/Y48W mutant, which has a substantially higher apparent $K_{0.5}$, however without concomitant changes in v_{max} (see Figure 4.5 legend). To demonstrate the specificity of the acidification, the assay was also carried out with uninjected oocytes and on wildtype H,K-ATPase-expressing oocytes in presence of 100 μM SCH 28080 (Figure 4.6 D).

Figure 4.7 summarizes the observed effects of the double tryptophan replacements on the E_1P/E_2P conformational distribution and the cation affinities of Na,K-ATPase (Figure 4.7 A and B) or H,K-ATPase (Figure 4.7 C and D). In both enzymes, the mutations result in a shift of the conformational equilibrium towards E_1P, as inferred from the voltage-dependent distribution of charge translocation and of fluorescence amplitudes (Figure 4.3 A and B for the sodium pump, Figure 4.5 A for the proton pump). Considering the voltage-dependent kinetics of charge translocation and fluorescence changes of the Na,K-ATPase $β_1$-(Y39W,Y43W) mutant compared to the wildtype (Figure 4.3 C and D), this can be assigned to an augmented apparent rate constant of Na^+ reverse binding (often designated as "backward" rate constant; k_{-1} in Figure 4.7), since the differences in the reciprocal time constants are most pronounced at extremely negative potentials. At positive potentials, however, no significant changes are observed, which suggests that the "forward" rate constant (k_1 in Figure 4.7) is unaltered by the mutation. Unfortunately, since the voltage dependence of the apparent rate constants is shallow for the proton pump and the fluorescence amplitudes for the H,K-ATPase β-(Y44W,Y48W) variant were quite small, it was not possible to demonstrate a similar phenomenon for the kinetics of the H,K-ATPase variant. Yet, since the Na,K- and H,K-ATPase variants are mechanistically very similar (as summarized in Figure 4.7), it is reasonable to assume that comparable k_1/k_{-1} alterations underlie the E_1P-shift of the H,K-ATPase mutant. Moreover, as the apparent v_{max} of Rb^+ uptake is unaffected for H,K-ATPase β-(Y44W,Y48W) (see Figure 4.5 legend), it is likely that only the backward rate constant k_{-1} is changed, since a decreased forward rate constant k_1 (which could as well explain an E_1-shift) would most likely also reduce the catalytic turnover number v_{max}. The shifted E_1P/E_2P distribution was shown to be accompanied by a decrease in the apparent affinity for extracellular K^+ of both enzymes, and enhanced rebinding of extracellular Na^+ for the sodium pump (or extracellular H^+ for the proton pump) was demonstrated for the variant pumps (Figure 4.7 B and D). For the Na,K ATPase, also an increased affinity for intracellular sodium was found (Hasler et al., 2001). However, whether an increased proton affinity at the intracellular binding sites of H,K-ATPase is caused by the β-(Y44W,Y48W) mutation, could not be determined by VCF experiments.

It should be emphasized that it is difficult to distinguish between cause and effect regarding the relation between shifts in conformational equilibria and changes in apparent cation affinities. If ion binding per se is disturbed, e.g. as a consequence of a mutation or reorientation of coordinating amino acids, this likely causes secondary shifts of the E_1P/E_2P distribution due to relative changes in k_1 or k_{-1}. However, regarding the β–TM mutants investigated here, we consider it more likely that the mutations introduced far from the site of cation coordination in the α-subunit have an destabilizing effect on the E_2or E_2P conformers of the holoenzyme. This in turn would result in a more effective ion binding to sites accessible in the E_1-state (intracellular Na^+) and a simultaneous decrease of ion binding to E_2-exposed sites (extracellular K^+ or Rb^+, but also their competitors Na^+or H^+).

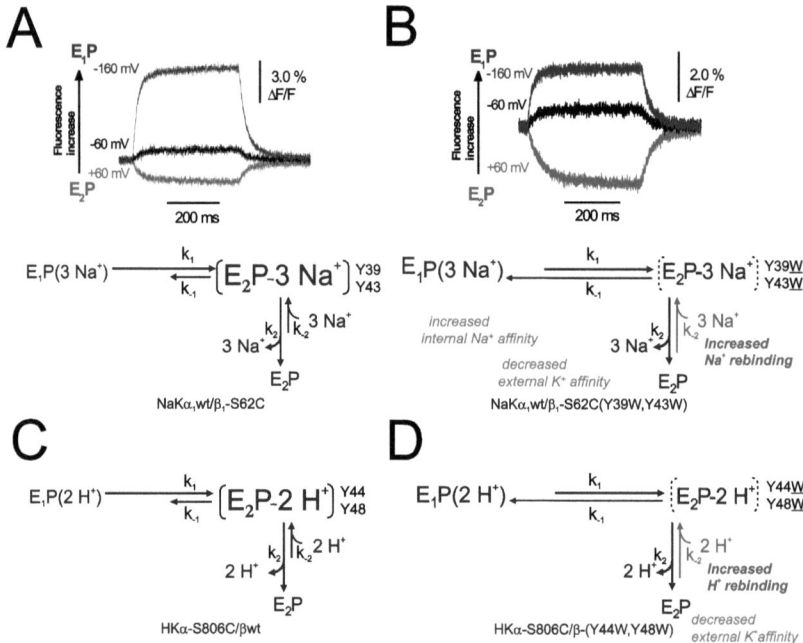

Figure 4.7.: **Schematic illustration of the putative changes in the E_1P/E_2P conformational and external ion binding equilibria caused by the tryptophan replacements in the Na,K-ATPase (A, B) and H,K-ATPase (C, D) β-TM.** According to this scheme the rate constants for the forward (k_1) and reverse (k_{-1}) $E_1P \rightarrow E_2P$ transitions are represented by arrows; k_2 and k_{-2} represent rate constants for extracellular release and reverse binding of Na^+ (A, B) or H^+ (C, D) respectively. Putative variations in these rate constants for the Na,K-ATPase β_1S62C(Y39W,Y43W) mutant (B) or the H,K-ATPase β(Y44W,Y48W) variant (D) from the "wildtype" rate constants (A,C) are indicated by a changed arrow length. The hypothetical stabilization of the E_2-state by the two tyrosines is symbolized by a parenthesis (A, C), which is disrupted for the mutant ATPases (B, D). For the Na,K-ATPase, representative fluorescence changes resulting from voltage-induced shifts of the E_1P/E_2P equilibrium are shown for the wildtype (A) and the β_1S62C(Y39W,Y43W) mutant (B) to demonstrate experimental support for the shifted E_1P/E_2P conformational distributions assumed here.

4.2.5. TM7 residues of Na,K- and H,K-ATPase α-subunits relevant for the E_2-stabilizing interaction with the two conserved β-tyrosines

According to the recently published crystal structures of the pig renal Na,K-ATPase in the Rb^+-occluded E_2-state (Morth et al., 2007) and the shark rectal gland Na,K-ATPase in a similar E_2-state with bound P_i (Shinoda et al., 2009), the two conserved β-tyrosines are at hydrogen bonding distance to TM7 of the α-subunit (Figure 4.1 B). Interestingly, TM7 is partially unwound at this putative interaction site around G848, resulting in a slight kink of the helix. Of note, the normally unfavorable unwinding is stabilized by the formation of a backbone hydrogen bond to Y43 in the β-TM. Since this unwinding was suggested to be of central importance for K^+ binding according to the crystal structure at 2.4 Å resolution (Shinoda et al., 2009), it corroborates the idea that the β-subunits of P_{2C}-type ATPases have evolved to enable the K^+ counter-transport activity exerted by these ion pumps (as described in section 1.1.4 on page 14).

Whereas in the H,K-ATPase sequence a phenylalanine (H,Kα-F864) is located in homologous position to this glycine (G848), the two adjacent residues Y847 and Q849 on the Na,K-ATPase α-subunit are conserved (Figure 4.1 D), corresponding to residues H,Kα-Y863 and H,Kα-Q865 respectively. Apart from the backbone amide oxygen of G848, the side chain of the conserved Q849 is a likely candidate to form hydrogen bonds with the hydroxyl groups (or amino-aromatic interactions with the phenyl moiety) of the two tyrosines (Figure 4.1 B). Of note, these hydrogen bonds are unique for the whole α/β transmembrane interaction interface (Figure 4.1 A), thus underlining their potential contribution to the overall stabilization of the conformational state. If the distances and geometry of the interacting residues were slightly altered in the E_1-state, a conformational change towards E_1 would involve an energetically disfavorable breakage of some of these H-bonds. Accordingly, a replacement of the two tyrosines by bulky tryptophans may disrupt this E_2-specific hydrogen bonding pattern and thus also the resulting E_2-stabilization, actually explaining the conformational shift towards the E_1-state observed for such mutant ATPases. If this is actually the case, tryptophan replacements of the potential interacting residues of the α-subunit should also impair this putative E_2-stabilization.

To test this hypothesis, we mutated the aforementioned three residues in the Na,K- and H,K-ATPase TM7 and determined the voltage-dependent E_1P/E_2P distributions for these α-subunit mutants when coexpressed with Na,K-ATPase $β_1$-S62C or wildtype H,K-ATPase β-subunits. The midpoint potentials of the resulting Q-V and (1-$\Delta F/F$)-V curves for the Na,K-ATPase variants are shown in Figure 4.8 A and B, respectively. Notably, tryptophan replacements of G848 and Q849 shifted the midpoint potentials of both (1-$\Delta F/F$) and Q-V distributions substantially towards more positive potentials. Although the observed shifts were less pronounced than for the $β_1$-(Y39W,Y43W) mutant, still the idea is supported that G848 is involved in an E_2-stabilizing interaction with the β-subunit. To analyze the functional relevance of H,Kα-F864 found instead of G848 in homologous position, we also investigated the Na,K-ATPase variant G848F. For this mutant, the voltage-dependent E_1P/E_2P distribution was similarly shifted towards E_1, thus indicating that the bulky phenylalanine side chain is as disruptive for the Na,K-ATPase α/β interaction as tryptophan. Both findings hint at an important role of the G848 backbone amide group for the interaction with the β-tyrosines, which is apparently disrupted when larger side chains are introduced.

Considering the Q849W mutant, a possible interpretation is that the observed E_1-shifted phenotype is directly caused by the removal of the glutamine side chain with its high propensity to participate in hydrogen bonding with β-Y39. On the other hand, the properties of Q849W could merely be a consequence of the bulkiness of the introduced tryptophan side chain, which interferes with the interaction mediated by the adjacent backbone amide of G848. In order to further clarify this issue, we also investigated the Q849G mutant which has no bulky side chain but cannot form side chain H-bonds, and the mutant Q849H, whose side chain is well suited to form hydrogen bonds. Of note, although Q849 is rather conserved among most α-subunits, a histidine is found in the corresponding position of the non gastric H,K-ATPase (Figure 4.1 D), thus possibly indicating a necessity for a hydrogen bonding side chain in this position. Keeping in mind that mainly the aromatic part and not the hydroxyl group of the β-tyrosines was crucial for the E_2-stabilizing effect (see section 4.2.2 on page 54), an amino-aromatic interaction to the positively charged or $δ^+$amino groups of glutamine or histidine is actually even more likely than "classical" hydrogen bonding: the side chain of Q849 (or H849) is within 6 Å of the ring centroid of β-Y39 (Figure 4.1 B) and thus optimally positioned to make van der Waals' contact with the $δ^-$ of the tyrosine's π-electrons according to Burley and Petsko (1986). This would also allow the formation of an amino-aromatic H-bond, as described between the side chain NH_2 of Asn-44 and Tyr-35 in bovine pancreatic trypsin inhibitor (Tüchsen and Woodward, 1987).

Yet, the midpoint potentials of Q849H and Q849G were not significantly different from the wildtype for both Q-V and (1-$\Delta F/F$)-V curves, which favours the idea that Q849 is not directly involved in the interaction to the β-subunit tyrosines, since its ability to form hydrogen bonds (or to provide $δ^+$amino groups for amino-aromatic interactions) does apparently not play a role for the E_1P/E_2P steady-state distribution. Therefore, rather the G848 backbone N-H is the prime candidate for donating an amino-aromatic hydrogen-bond to the phenyl rings of β-Y39 or β-Y43, since the centroids of both are within the aforementioned critical distance of 6 Å.

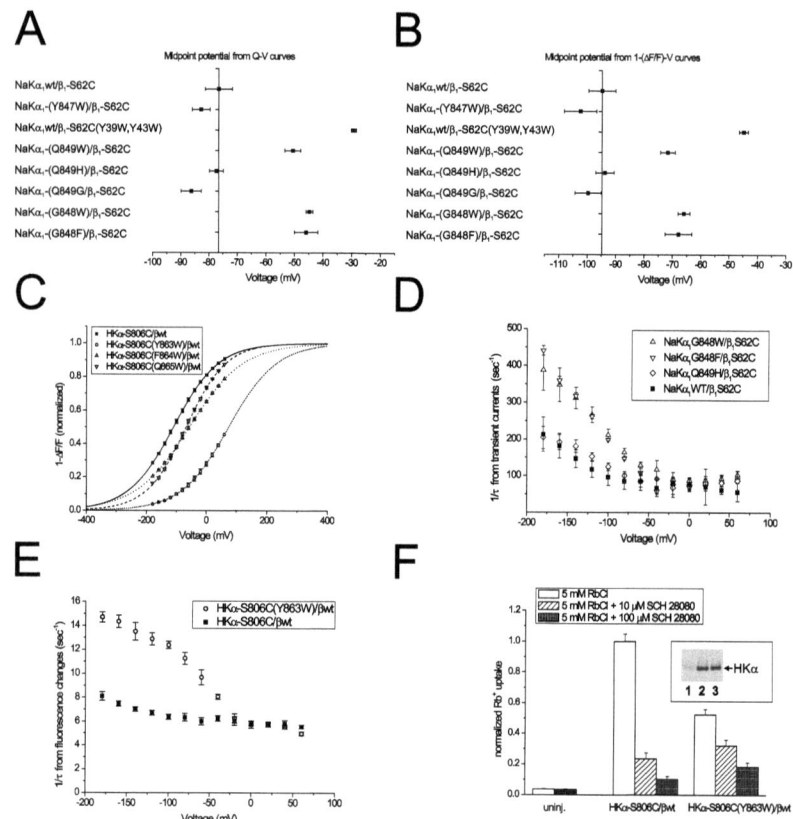

Figure 4.8.: **Voltage-dependent E_1P/E_2P distribution of Na,K- and H,K-ATPase α-TM7 variants.** A, B: Midpoint potentials derived from fits of a Boltzmann function to the voltage-dependent distributions of transported charge Q (A) or of fluorescence amplitudes 1-$\Delta F/F$ (B) for Na,K-ATPase complexes consisting of either the wildtype α_1-subunit or various α_1-TM7 mutants co-expressed with construct β_1-S62C. C: Voltage-dependent E_1P/E_2P distribution of fluorescently labeled H,K-ATPase αS806C/βwt and three α-TM7 variants: HKαS806C(Y863W)/βwt, HKαS806C(F864W)/βwt and HK06C(Q865W)/βwt. D: Reciprocal time constants of voltage-jump induced transient currents for oocytes expressing either the wildtype Na,K-ATPase α_1-subunit or various α_1-TM7 mutants together with reporter construct β_1S62C. E: Reciprocal time constants of voltage-jump induced fluorescence changes for oocytes expressing the H,K-ATPase wildtype β-subunit together with either H,K-ATPase HKαS806C or HKαS806C(Y863W). F: Rb^+ uptake of uninjected oocytes and oocytes expressing either HKαS806C/βwt or HKαS806C(Y863W)/βwt. Rb^+ uptake was determined at pH 5.5 in extracellular Na^+-free solution containing 5 mM RbCl. Data are means ± S.E., n = 50-60 oocytes from 4 independent experiments. Inset: Western blot analysis of isolated plasma membranes from uninjected oocytes (1) or oocytes expressing either HKαS806C/βwt (2) or HKαS806C(Y863W)/βwt (3), using an anti-H,Kα antibody HK12.18. All electrophysiological data are means ± S.E. of 10-20 oocytes from 2-3 different *Xenopus* females.

construct	Boltzmann parameter (1-ΔF/F)-V curves		Boltzmann parameter Q-V curves	
	$V_{0.5}$	z_q	$V_{0.5}$	z_q
Na,K$\alpha_1$wt/β_1S62C	-94.8 ± 4.8	0.74 ± 0.01	-76.6 ± 4.8	0.77 ± 0.03
Na,Kα_1-Y847W/β_1S62C	-102.4 ± 5.7	0.56 ± 0.04	-82.8 ± 3.2	0.56 ± 0.03
Na,Kα_1-G848W/β_1S62C	-66.0 ± 2.1	0.70 ± 0.02	-44.8 ± 1.2	0.76 ± 0.01
Na,Kα_1-G848F/β_1S62C	-67.9 ± 4.7	0.74 ± 0.01	-45.9 ± 4.1	0.76 ± 0.02
Na,Kα_1-Q849W/β_1S62C	-71.6 ± 2.6	0.75 ± 0.05	-50.6 ± 2.8	0.76 ± 0.03
Na,Kα_1-Q849G/β_1S62C	-99.7± 4.7	0.72 ± 0.02	-86.3 ± 3.6	0.66 ± 0.02
Na,Kα_1-Q849H/β_1S62C	-93.8 ± 3.2	0.72 ± 0.01	-77.4 ± 2.5	0.75 ± 0.01
H,KαS806C/βwt	-110.8 ± 8.6	0.34 ± 0.01	-	-
H,KαS806C(Y863W)/βwt	+74.3 ± 12.7	0.32 ± 0.02	-	-
H,KαS806C(F864W)/βwt	-58.0 ± 7.0	0.28 ± 0.02	-	-
H,KαS806C(Q865W)/βwt	-67.9 ± 3.6	0.36 ± 0.01	-	-

Table 4.2.: **Parameters from fits of a Boltzmann function to (1-ΔF/F)-V or Q-V distributions of Na,K-ATPase or H,K-ATPase α-TM7-variants.** Values are means ± S.E. of 15-25 oocytes from 2-3 oocyte batches.

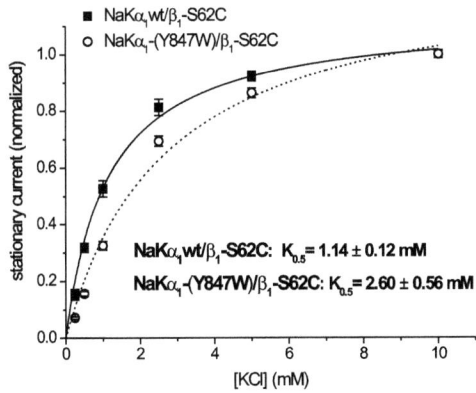

Figure 4.9.: **Apparent affinity for extracellular K^+ of the Na,K-ATPase α_1-TM7 mutant Y847W.** K^+-activated pump currents were determined at a holding potential of -40 mV. Data are means ± S.E. of 10-15 oocytes. The apparent $K_{0.5}$ were obtained by fitting of a Michaelis-Menten-type function to the data.

Interestingly, tryptophan replacement of Y847 results in Q-V and (1-ΔF/F)-V curves with unaltered $V_{0.5}$ values but significantly reduced slope factors z_q (0.56 instead of ~0.75 for the wildtype, see Table 4.2). Apparently this residue is not relevant for the supposed E_2-specific β-interaction, which seems reasonable since its side chain is not directed toward the β-subunit but rather points toward the cation binding pocket (Figure 4.1 B). According to the recently published Na,K-ATPase structure at 2.4 Å resolution (Shinoda et al., 2009), the side chain of Y847 (corresponding to Y854 in the shark ATPase) forms a hydrogen-bond to N776 (corresponding to N783 in the shark structure), which is a critical part of the K^+ binding sites I and II. In fact, this interaction was suggested to be a key player in defining K^+ specificity of the cation binding site, since it tethers the N776 side-chain into a distinct position which is markedly different from the corresponding Ca^{2+}-coordinating N768 residue of the Ca^{2+}-ATPase. Notably, the authors expected that a disruption of this interaction occurs upon the

construct	apparent $K_{0.5}(Rb^+)$
H,KαS806C/βwt	0.26 ± 0.03
H,KαS806C(Y863W)/βwt	0.22 ± 0.05
H,KαS806C(F864W)/βwt	0.24 ± 0.05
H,KαS806C(F864G)/βwt	0.25 ± 0.04
H,KαS806C(Q865W)/βwt	0.20 ± 0.03
H,KαS806C(Q865G)/βwt	0.23 ± 0.05
H,KαS806C(Q865H)/βwt	0.25 ± 0.07

Table 4.3.: **Apparent $K_{0.5}$ values for halfmaximal Rb^+ activation of H,K-ATPase α-TM7-variants.** Rb^+ uptake measurements at various extracellular Rb^+ concentrations were performed in absence of extracellular Na^+ at pH 5.5. Values are means ± S.E. of at least two independent experiments.

$E_2 \rightarrow E_1$ conformational transition, thus probably adjusting the N776 side-chain for coordination of the smaller Na^+ ions in the E_1-state. Keeping in mind that electrogenicity of the sodium pump was mainly linked to Na^+ binding and -release steps (Chapter 1, section 1.1.2), this might also explain why a parameter is altered (lowered z_q) which reflects voltage-dependent properties of the enzyme. The reduced z_q value can be interpreted in terms of a reduced apparent depth of the proposed ion access channel for extracellular Na^+ reverse binding or release (Sagar and Rakowski, 1994), as a consequence of the tryptophan replacement. Moreover, we determined the apparent affinity for extracellular K^+ by measuring the K^+-dependence of pump current amplitudes and found a significantly higher apparent $K_{0.5}$ for this mutant (2.6 mM instead of 1.1 mM for the wildtype at -40 mV, see Figure 4.9), which confirms that Y847 plays indeed a critical role for K^+ coordination of the extracellular cation binding sites.

Regarding the H,K-ATPase α-subunit variants, a pronounced positive shift of the $(1-\Delta F/F)$-V curves was observed for all three investigated variants H,Kα-Y863W, -F864W and -Q865W (Figure 4.8 C). Surprisingly, the most pronounced effect was observed for the Y863W mutant, which shifted the midpoint potential of the E_1P/E_2P distribution by about +180 mV towards extremely positive potentials.

Notably, the kinetics of transient currents of the E_1-shifted Na,K-ATPase α-TM7 mutants Na,Kα-G848W and Na,Kα-G848F were substantially faster than the wildtype (or the Na,Kα-G848H mutant which showed no changes in $V_{0.5}$) at hyperpolarizing potentials, but not at depolarizing potentials (Figure 4.8 D). Likewise, the reciprocal time constants obtained from monoexponential fitting to the fluorescence changes were almost increased two-fold for the strongly E_1-shifted H,K-ATPase α-TM7 mutant H,Kα-Y863W compared to the wildtype at very negative potentials, whereas there was no significant difference at positive potentials (Figure 4.8 E). This can be interpreted as an acceleration of the backward rate constant k_{-1} (Figure 4.7) without concomitant changes in k_1, thus resulting in a E_1-shift of the E_1P/E_2P conformational distribution, very similar to the Na,K-ATPase β_1-S62C(Y39W,Y43W) variant (Figure 4.3 C, D). Again, the E_1P-shift is also reflected by a reduced SCH 28080 sensitivity of the Y863W mutant compared to the HKαS806C reference construct in presence of 5 mM RbCl and 10 or 100 μM SCH 28080, respectively (Figure 4.8 F). Moreover, the Rb^+ uptake of the H,Kα-Y863W variant was reduced to about 50% of the wildtype uptake at 5 mM RbCl (Figure 4.8 F), but this could not be attributed to changes in the apparent Rb^+ affinity (Table 4.3) or a reduced cell surface delivery (Figure 4.8 F, inset). Therefore, the turnover number of the mutant enzyme is apparently affected, too.

The finding that the most pronounced effects of tryptophan replacements in the H,K-ATPase TM7 occured in position Y863 (and not F864, corresponding to NaK-G848) indicates that the location of the interaction interface on the α-subunit might be different for H,K- and Na,K-ATPase enzymes, since the corresponding residue Y847 of the sodium pump did not seem important for the E_1P/E_2P distribution. Furthermore, even the mechanism of the interaction may be different for the two ATPases: The side chains of Y863 and F864 in the H,K-ATPase TM7 probably exert their E_2-stabilizing effect rather by aromatic-aromatic interactions (π-stacking, see Burley and Petsko (1985)) with Y44 and Y48 in

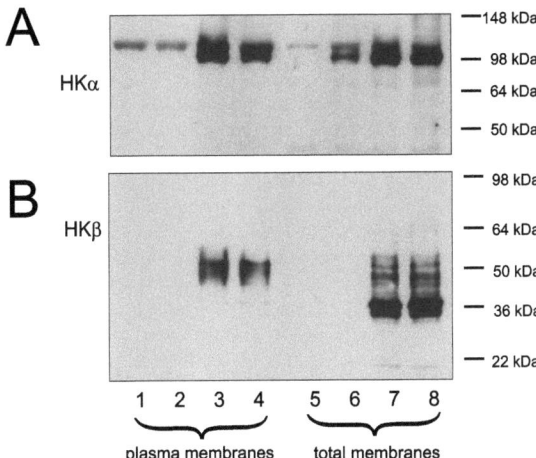

Figure 4.10.: **Plasma membrane delivery of the H,K-ATPase β-(Y44W,Y48W) mutant.**
Plasma membrane (1-4) and total membrane (5-8) fractions from oocytes injected with cRNAs for HKαS806C only (lanes 2 and 6), HKαS806C + HKβwt (lanes 3 and 7), or HKαS806C + HKβ-(Y44W,Y48W) (lanes 4 and 8), or from uninjected oocytes (lanes 1 and 5). The equivalent of two oocytes was probed by either α-specific antibody HK12.18 (A) or β-specific antibodiy 2G11 (B), respectively.

the β–TM. Yet, despite the huge effects on the E_1P/E_2P conformational equilibrium, none of these H,K-ATPase mutations exhibited altered apparent affinities for extracellular Rb^+ (see Table 4.3). Therefore, a similar prominent role of this conserved tyrosine in K^+ coordination (as suggested for the sodium pump by the recent 2.4 Å crystal structure, see above) seems rather unlikely for the gastric H,K-ATPase.

This apparently contradicts findings obtained for the β-(Y44W,Y48W) mutant, which showed a less pronounced E_1-shift of the $(1-\Delta F/F)$-V curve, but exhibited a substantially reduced apparent affinity for extracellular Rb^+. However, the VCF data for this β-subunit mutant should be considered cautiously since the observed fluorescence changes were only about 10% compared to those obtained for wildtype or mutant α-subunit H,K-ATPases. The unaffected maximal Rb^+ uptake per cell at saturating Rb^+ concentrations (see legend to Figure 4.5 B, C) indicates that the lower fluorescence amplitudes are not due to a reduced cell surface delivery of the β-(Y44W,Y48W) mutant. This interpretation was also confirmed by Western blots of isolated plasma membrane protein fractions (see Figure 4.10), which show that the amounts of protein at the cell surface are very similar. Therefore, we suggest instead that the β-(Y44W,Y48W) mutant has an extremely high preference for the E_1P-state (as inferred from the dramatic Rb^+ affinity changes and the reduced SCH 28080 sensitivity), which prevents that the enzymes can completely be shifted towards E_2P by stepping to positive potentials. Since only relative changes in the conformational distribution are reported by the VCF technique, the weak fluorescence signals observed for the β-(Y44W,Y48W) variant enzyme indicate that most likely an incomplete $E_1P \rightarrow E_2P$ conformational transition is monitored, thus the absolute E_1P-shift for the mutant enzyme might be underestimated.

4.2.6. Physiological relevance of the β-subunits' E_2-stabilizing effect for Na,K- and gastric H,K-ATPase activity *in situ*

In this chapter, a common E_2-stabilizing effect of conserved Tyr residues in the β-subunit TM for oligomeric Na,K- and H,K-ATPase was demonstrated and insights into the structural determinants of the E_2-specific interaction underlying this stabilization were provided. The functional significance is less pronounced for the Na,K-ATPase, where an E_1-shifted equilibrium caused by the TM mutations did only involve minor changes in the apparent ion affinities, such that no substantial effect on the transport rate of the sodium pump is exerted under physiological conditions.

In contrast, the E_1-preference caused by the homologous β-(Y44W,Y48W) mutation in gastric H,K-ATPase has serious effects on cation affinities: Since the apparent $K_{0.5}$ for extracellular K^+ (Rb^+) was shown to be more than 4-fold increased (in presence of extracellular Na^+), the mutated H,K-ATPase is not able to operate at maximal turnover *in situ*. Usually, luminal K^+ concentrations are low, at least before K^+-secreting KCNQ1/KCNE2 potassium channels in the luminal parietal cell membrane are activated by acidification to about pH 3.5 (Grahammer et al., 2001; Heitzmann et al., 2004), a process which itself requires H,K-ATPase activity (see also chapter 1, section 1.2.2.2 on page 22). Moreover, even an increase of the potassium concentration in the stomach lumen to saturating levels of about 15 mM (Sachs et al., 2007), e.g. by K^+-rich diet or as a result of sustained KCNQ1/KCNE2 activity, would not help much, since the enhanced H^+ rebinding of the mutant would still be fatal for efficient H^+ secretion (as demonstrated at 20 mM KCl, Figure 4.6). For the H,K-ATPase it is of pivotal importance that protons can be effectively released from the extracellular binding sites, even at pH-values as low as 1. Since this requires pK_a changes of 5 to 6 orders of magnitude during the $E_1P \rightarrow E_2P$ transition, the relative destabilization of the E_2-state in favour of E_1, which is characteristic for the mutant, most likely results in a substantially lowered efficiency of luminal proton release. In the stomach, the combination of both, decreased apparent K^+ affinity and reduced proton release, is expected to suppress H,K-ATPase activity dramatically, thus rendering the mutant enzyme unable to sustain sufficiently low pH-levels required for digestion and antibacterial purposes. Despite the observed unaffected Rb^+ affinity, the phenotype resulting from the H,Kα-Y863W mutation would also result in a significant reduction of H^+secretion in the stomach, since the turnover number (v_{max}) of this mutant is about two-fold lowered. Therefore, the E_2-stabilization mediated by E_2-specific intersubunit interactions between two conserved tyrosines in the Na,K- or H,K-ATPase β-TM and TM7 of the respective catalytic α-subunit, which is the main finding of this study, is of high functional significance for the gastric H,K-ATPase *in situ*.

CHAPTER 5

E_2P-state stabilization by the N-terminal tail of the H,K-ATPase β-subunit is critical for efficient proton pumping under *in vivo* conditions

Dürr et al. (2009) The Journal of Biological Chemistry 284, 20147–20154

5.1. Introduction

It was already mentioned in the previous chapters that compared to the extensively studied Na,K-ATPase, far less is known about the H,K-ATPase β-subunit 's influence on the enzyme's cation transport. Whereas some experimental support for a possible contribution of the β-subunit's ectodomain (Chow et al., 1992) and TM region (see previous chapter 4) to ion transport activity of the H,K-ATPase already exists, not much is known so far about the functional significance of the short N-terminal tail.

In studies on the Na,K-ATPase, truncation of the whole N-terminal domain of the β-subunit had no effect on α/β coassembly and cell surface expression (Renaud et al., 1991; Hasler et al., 1998), but profoundly affected the ion translocating function of the holoenzyme, including the apparent affinities for extracellular K^+ and intracellular Na^+ (Geering et al., 1996; Shainskaya and Karlish, 1996; Hasler et al., 1998). Moreover, the distribution between E_1P/E_2P-states was strongly influenced by deletion of the N-terminal domain (Abriel et al., 1999). Yet, the most striking finding was an ouabain-sensitive inward current in the absence of K^+, which was presumably driven by protons. The authors suggested that the lack of the β-N-terminus favours the formation of a "leaky" H^+-conductive E_2P-state, even in presence of extracellular high $[Na^+]$ (Abriel et al., 1999). However, shorter deletions or multiple mutations in the β-N-terminus had no effect on the K^+-induced pump currents of the Na,K-ATPase.

Interestingly, it was shown that the H,K-ATPase β-N-terminus can effectively substitute for that of the Na,K-ATPase, resulting in a chimeric β-subunit that retains ATPase activity and produces K^+-stimulated pump currents upon coexpression with the Na,K-ATPase α-subunit (Ueno et al., 1997). Vice versa, ATPase activity was also unaffected when the H,K-ATPase α-subunit was co-expressed with a chimeric β-subunit consisting of the H,K-ATPase β-ectodomain and the cytoplasmic plus transmembrane domains of the Na,K-ATPase β-subunit (Asano et al., 1999). This may on the one hand imply that the cytoplasmic N-terminal domain derived from the Na,K-ATPase β-subunit can effectively substitute for the N-terminus of the H,K-ATPase β-subunit. On the other hand, measurement of ATPase activity alone neither provides information about a potential contribution of the β-subunit's N-terminal tail to ion transport nor about possible shifts of conformational equilibria.

However, one possible hint that the β-N-terminus may be important for ion transport of the gastric H,K-ATPase stems from the inhibitory action of the monoclonal antibody Mab2G11, which recognizes the 36 N-terminal residues of H,K-ATPase β-subunit (Chow and Forte, 1993). Binding of Mab2G11 was shown to affect H,K-ATPase activity by interfering with the conformational transition induced by K^+.

Furthermore, according to the recently published cryo-EM structure of pig gastric H,K-ATPase in the pseudo E_2P-state (see section 1.1.3 on page 14), the N-terminal tail of the β-subunit is in close proximity to the phosphorylation domain (P-domain) of the α-subunit (indicated by the red arrow in Figure 1.5 on page 13). The authors suggested that a contact between the two subunits could possibly stabilize the enzyme in the E_2P-state, thereby facilitating efficient proton release against the extraordinarily steep H^+ gradients in the stomach (Abe et al., 2009).

To test this hypothesis, the experimental approaches from the previous chapters were applied to characterize several N-terminally truncated H,K β-variants (see Figure 5.1) regarding their distribution between E_1P/E_2P-states and the kinetics of the respective conformational transition. Furthermore, to assess the functional consequences of a potential E_2-stabilizing effect mediated by the β-N-terminus on ion transport activity of the gastric H,K-ATPase, the concentration-dependent Rb^+ uptake of these mutants was also studied.

5.2. Results

5.2.1. Cell surface expression of N-terminally deleted H,K β-variants.

To first determine whether N-terminal deletions already impair the β-subunit's chaperone-like functions, such as α/β coassembly or plasma membrane targeting of the H,K-ATPase, plasma membranes were isolated from *Xenopus* oocytes and subjected to Western blot analysis. According to Figure 5.2, all N-terminally deleted β-mutants were not detected by the H,K-ATPase β-subunit specific antibody 2G11, since it recognizes an epitope located to the 36 N-terminal residues of the H,K-ATPase wildtype

```
HKβWT   MAALQEKKSCSQRMAEFRQYCWNPDTGQMLGRTPARWVWTM
HKβΔ4      MQEKKSCSQRMAEFRQYCWNPDTGQMLGRTPARWVWTM
HKβΔ8         MSCSQRMAEFRQYCWNPDTGQMLGRTPARWVWTM
HKβΔ13              MAEFRQYCWNPDTGQMLGRTPARWVWTM
HKβΔ29                         MLGRTPARWVWTM
        TM = ISLYYAAFYVVMTGLFALCIYVLMQTI
```

Figure 5.1.: **Partial amino acid sequence of the wildtype and N-terminally truncated mutant β-subunits of rat gastric H,K-ATPase.**

β-subunit (Figure 5.2 B, lane 2). Apparently, a deletion of the three very N-terminal residues (βΔ4) is already sufficient to prevent antibody binding. However, if antibody 2B6, which recognizes amino acids 236-281 in the extracellular C-terminus of the H,K-ATPase β-subunit is used instead, similar amounts of the N-terminal deleted β-variants are detected in both plasma- and total cellular membranes (Figure 5.2 C). Hence, cell surface expression of the β-subunit appears to be unaffected by the N-terminal truncations.

This can be additionally concluded from the high-level cell surface delivery of H,K-ATPase α-subunits which is facilitated by coexpression of the variant β-subunits: As shown in Figure 5.2 A (upper panel) the amount of H,K-ATPase α-subunit in the plasma membrane was the same for oocytes expressing βWT or any of the N-terminally deleted β-variants. Moreover, the amount of protein detected by the α-subunit-specific antibody in the total cellular membrane fraction (lower panel in Figure 5.2 A) was also identical to βWT-expressing oocytes. Together, these results indicate that the β-N-terminal truncations affect neither the efficiency of the β-variants to associate with H,K-ATPase α-subunits, nor plasma membrane targeting of the oligomeric enzyme.

5.2.2. E_1P/E_2P conformational distribution of N-terminally deleted H,K-ATPase β-mutants.

The wildtype β-subunit and the aforementioned N-terminally deleted H,K β-mutants were coexpressed with the α-subunit variant S806C in *Xenopus* oocytes, and voltage-jump-induced fluorescence changes upon site-specific labeling with TMRM were recorded (see Figure 5.3 A). Figure 5.3 B shows the resulting voltage-dependence of steady-state fluorescence amplitudes (1-$\Delta F/F$), which follow a Boltzmann-type distribution for H,K-ATPase enzymes assembled from wildtype as well as N-terminally deleted β-variants. Whereas these were not changed for the shorter N-terminal deletions Δ4 and Δ8 compared to the wildtype, the distributions for the βΔ13 and βΔ29 truncated variants were significantly shifted towards more positive potentials (see Table 5.1 for Boltzmann parameters), indicating a relative destabilization of the E_2P-state in favor of E_1P. According to these distributions, about 65% of the βWT, βΔ4 and βΔ8-containing H,K-ATPase molecules occur in the E_2P-state at -60 mV (which is the physiologically relevant membrane potential determined for parietal cells (Demarest and Machen, 1985)), but only 60% and 50% of the βΔ29 and βΔ13 mutant enzymes are present in E_2P, respectively. Furthermore, the E_1P-shift is also reflected by the voltage-dependent reciprocal time constants obtained from monoexponential fits to the fluorescence changes (Figure 5.3 C): As these are significantly smaller for the Δ29 and Δ13 mutants than for the wildtype only at positive potentials, which favor the transition to E_2P, but not at negative potentials, at which the enzyme is driven into the E_1P-state, these N-terminal deletions apparently reduce the "forward" rate constant (designated as k_{fw}) of the $E_1P \rightarrow E_2P$ transition without changing the rate constant k_{rev} for the "reverse" reaction. (Remarkably, these phenotypes observed for the βΔ13 and βΔ29 mutants clearly differ from those observed for the E_1P-shifted βTM- and αTM7 mutants described in the previous chapter 4, since the latter mutants showed an increased k_{rev} without concomitant changes in k_{fw}.)

However, if only the results shown in Fig. 5.3 B and C are considered, the conformational effect caused by the N-terminal deletions is probably underestimated, since the two variants βΔ13 and βΔ29

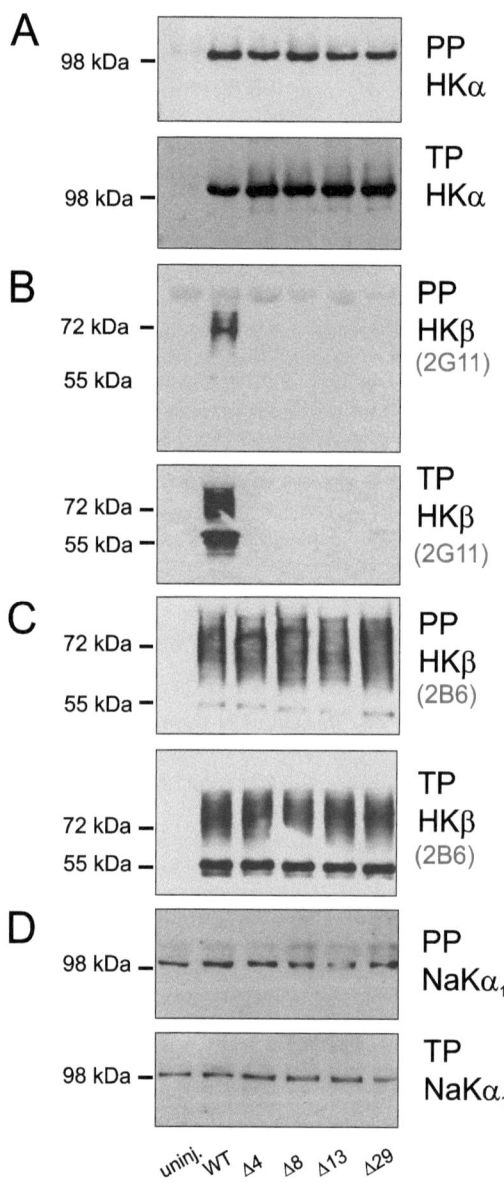

Figure 5.2.: **Western Blot analysis of N-terminally-deleted H,K-ATPase β-subunit variants.** Oocytes were uninjected (lane 1), or injected with cRNAs for HKα-S806C and cRNAs of either HKβwt (lane 2) or N-terminal deleted HK β-variants HKβΔ4 (lane 3), HKβΔ8 (lane 4), HKβΔ13 (lane 5) and HKβΔ29 (lane 6), respectively. The equivalent of two oocytes from plasma membrane (PP) or total membrane (TP) fractions was probed by either the α-specific antibody HK12.18 (A) or the β-specific antibodies 2G11 (B) and 2B6 (C). Detection of the endogenous *Xenopus* Na,K-ATPase α_1-isoform by antibody C356-M09 served as a loading standard (D).

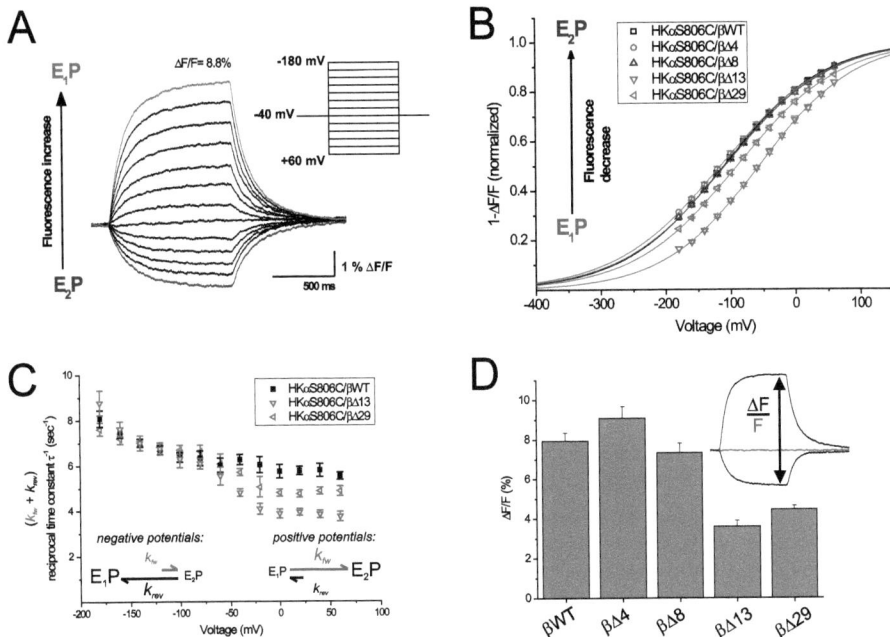

Figure 5.3.: **Voltage-dependent E_1P/E_2P distribution and kinetics of the $E_1P \rightarrow E_2P$ transition of N-terminally truncated H,K-ATPase β-variants.** A, Voltage-pulse induced fluorescence signals of a TMRM-labeled oocyte expressing H,K-ATPase α-S806C/βwt under high Na^+/K^+-free conditions at pH 5.5. Inset: voltage-protocol. B, Voltage-dependent distributions of fluorescence amplitudes 1-ΔF/F of wildtype or N-terminally deleted H,K-ATPase β-variants, normalized to saturating values obtained from fits of a Boltzmann function to the data. C, reciprocal time constants of voltage jump-induced fluorescence changes for wildtype and N-terminally deleted H,K-ATPase β-variants. Data are means ± S.E. of 10-16 oocytes. D, ΔF/F values (in %) for TMRM-labeled oocytes expressing α-S806C and either the wildtype H,K-ATPase β-subunit or N-terminally truncated β-variants. Data are means ± S.E. of 13-32 oocytes from 3 different oocyte batches. Inset: ΔF/F is calculated from the difference ΔF (represented by a black arrow) between the fluorescence at the most hyperpolarizing voltage (-180 mV) and the most depolarizing voltage (+60 mV), normalized to the background fluorescence F at -40 mV.

also showed substantially smaller ΔF/F values (Figure 5.3 D). The previously described Western blot analysis of isolated plasma membranes exclude the possibility that the lower ΔF/F values observed for the βΔ13 and βΔ29 constructs were simply due to a reduced cell surface delivery of the α-subunits that are available for labeling with TMRM (see previous subsection). Therefore, the shown ΔF/F ratio is directly proportional to the number of H,K-ATPase molecules that can be shifted between E_1P/E_2P-states by voltage-jumps. The approximately two-fold lower ΔF/F values observed for the βΔ13 and βΔ29 mutants (compared to βWT, βΔ4 and βΔ8) most likely reflect a higher tendency of the mutant H,K-ATPase molecules to undergo the backward reaction sequence E_1P $(H^+) \rightarrow E_1P$ + H^+ + ADP $\rightarrow E_1$ + H^+ + ATP and accumulate in the E_1P-state according to the reduced k_{fw}. This would result in a substantial depletion of the sum of E_1P/E_2P-states, which could explain the observed lower fluorescence changes. This interpretation is actually in line with previous results on N-terminally deleted mutants βΔ8, βΔ13 (Abe et al., 2009) or βΔ28 (K. Abe, unpublished results),

construct	Boltzmann parameter (1-$\Delta F/F$)-V curves	
	$V_{0.5}$	z_q
H,KαS806C/βwt	-110.1± 5.0	0.33 ± 0.01
H,KαS806C/$\beta\Delta 4$	-115.3± 5.8	0.31 ± 0.01
H,KαS806C/$\beta\Delta 4$	-115.3± 5.8	0.31 ± 0.01
H,KαS806C/$\beta\Delta 8$	-108.7 ± 4.3	0.32 ± 0.02
H,KαS806C/$\beta\Delta 13$	-56.3 ±6.3	0.34 ± 0.02
H,KαS806C/$\beta\Delta 29$	-89.9 ± 4.3	0.32 ± 0.01

Table 5.1.: **Parameters from fits of a Boltzmann function to (1-ΔF/F)-V distributions of N-terminally deleted H,K-ATPase β-mutants.** Values are means ± S.E. of 15-30 oocytes from 2-3 oocyte batches.

showing increased reactivity of radiolabeled $E_1{}^{32}P$ phosphoenzymes with ADP to form $\gamma^{32}P$-ATP via the aforementioned reverse reaction.

5.2.3. Rb$^+$uptake kinetics and SCH 28080 sensitivity of N-terminally truncated H,K-ATPase β-subunits

Rb$^+$ uptake measurements were performed to assess the potential impact on ion transport activity by the conformational shift observed for the N-terminally truncated β-variants. Whereas Rb$^+$ uptake at saturating extracellular concentrations was comparable to WT for $\Delta 4$- and $\Delta 8$-expressing oocytes, it was about 20% lower for $\Delta 13$ and $\Delta 29$ (Fig.5.4 A, Table 5.2). This can be interpreted as a reduced turnover number (lowered v_{max}), since the apparent affinity for extracellular Rb$^+$was unaffected by the truncations (see Figure 5.4 B and Table 5.2 for apparent $K_{0.5}$ values). The affected turnover numbers of the $\Delta 13$ and $\Delta 29$ mutants demonstrate that obviously the transition which is characterized by the reduced rate constant k_{fw} of these variants directly affects the rate-limiting step of the catalytic cycle. This illustrates how already small shifts in conformational equilibria can have significant functional consequences for the transport activity of gastric H,K-ATPase.

In presence of 10 µM SCH 28080, an E_2-specific inhibitor of the H,K-ATPase (see chapter 1, section 1.2.2.3, Figure on page 27), the Rb$^+$ uptake of WT, $\Delta 4$- and $\Delta 8$-expressing oocytes was reduced to approximately 20% (Figure 5.4A inset, Table 5.2), in agreement with the data of Mathews et al. (1995). In contrast, the inhibition of ATPase complexes containing $\Delta 13$ and $\Delta 29$ was less efficient, resulting in significantly higher residual activities of about 30% at 10 µM SCH 28080. Notably, 100 µM SCH 28080 resulted in a suppression of Rb$^+$ uptake to about 6% for H,KαS806C/βwt, but for any of the N-terminally deleted variants the effect was much smaller. The dwell time of molecules in the SCH 28080-sensitive E_2-state may be substantially lower for the E_1P-shifted mutants, so that increasing the inhibitor concentration will not result in enhanced binding of the compound, since the dwell time in E_2 is not sufficient to reach binding equilibrium. In contrast, the wildtype H,K-ATPase, which stays longer in E_2, is able to bind more inhibitor molecules if the SCH 28080 concentration is increased. Notably, at the higher inhibitor concentration, a significantly reduced SCH 28080 sensitivity was also observed for $\Delta 4$- and $\Delta 8$-expressing oocytes, which showed a more than two-fold higher residual activity compared to WT (13-14% versus 6%).

This suggests that already the shorter deletions cause an elevated preference for E_1, which raises the question, why no effect on the conformational distribution (data in Figure 5.3 B) was seen for these constructs. Two possibilities may account for this: i) minute shifts in the voltage-dependent E_1P/E_2P distribution may be difficult to resolve by VCF experiments, since low slope factor z_q of the Boltzmann curves limits the accuracy of $V_{0.5}$ determination. ii) E_2-destabilizing effects that act on the relative distribution of pump molecules over all reaction intermediates may not be effective during the partial reaction sequence studied in pre-steady-state experiments (VCF), but rather become apparent under steady-state conditions (Rb$^+$ uptake), in which the enzyme undergoes the full reaction cycle.

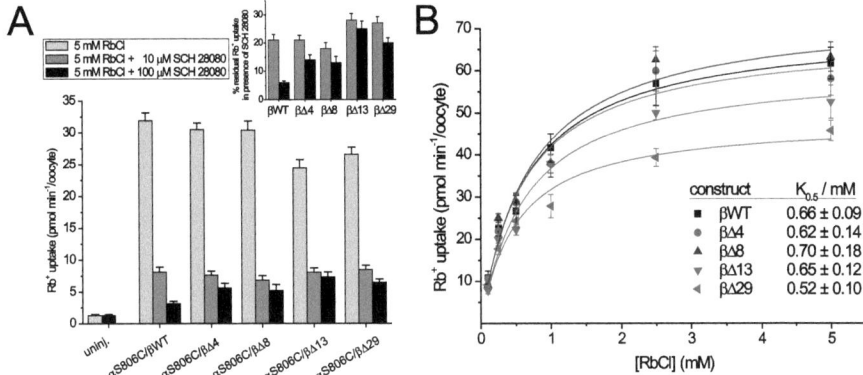

Figure 5.4.: **Rb^+ uptake, SCH 28080 sensitivity and apparent Rb^+ affinity of β-N-terminally truncated H,K-variants.** A, Rb^+ uptake at pH 5.5 in extracellular Na^+-free solutions containing 5 mM RbCl in absence (light grey bars) or presence of SCH 28080 (dark grey bars: 10 μM, black bars: 100 μM). Inset: Residual Rb^+ uptake activity (in %) in presence of 10 μM (dark grey bars) or 100 μM (black bars) SCH 28080. Data were normalized to Rb^+ uptake at 5 mM RbCl in absence of SCH 28080 for each construct after subtraction of the mean Rb^+ uptake of uninjected oocytes. B, Michaelis-Menten plots for concentration dependent Rb^+ uptake of wildtype or N-terminally deleted H,K-ATPase β-variants.

construct	Normalized Rb^+ uptake activity (in %)[a,b] at 5 mM RbCl	Residual Rb^+ uptake activity in presence of SCH 28080 (in %)[a,c]		apparent Rb^+ affinity[d] $K_{0.5}$(mM)
		10 μM SCH 28080	100 μM SCH 28080	
H,KαS806C/βwt	100 ± 3.9	21 ± 2.0	6 ± 0.7	0.66 ± 0.09
H,KαS806C/βΔ4	96 ± 3.3	21 ± 1.7	14 ± 1.8	0.62 ± 0.14
H,KαS806C/βΔ8	95 ± 4.4	18 ± 2.0	13 ± 2.3	0.70 ± 0.18
H,KαS806C/βΔ13	77 ± 4.0	28 ± 2.3	25 ± 2.6	0.65 ± 0.12
H,KαS806C/βΔ29	83 ± 3.8	27 ± 2.2	20 ± 1.7	0.52 ± 0.10
uninjected	4 ± 0.4	-	-	-

[a]Values are means ± S.E. of 40-50 oocytes from 3-4 different oocyte batches
[b]Data were normalized to Rb^+ uptake of H,KαS806C/βwt, corresponding to 32 pmol min^{-1}/oocyte
[c]Data were normalized to Rb^+ uptake at 5 mM RbCl in absence of SCH 28080 for each construct after subtraction of the mean Rb^+ uptake of uninjected oocytes
[d]Values are means ± S.E. of 10-20 oocytes

Table 5.2.: **Normalized values for Rb^+ uptake, SCH 28080 sensitivity and apparent Rb^+ affinity of β-N-terminally truncated H,K-ATPase variants.**

5.3. Discussion

5.3.1. Physiological significance of the β-N-terminus for gastric H,K-ATPase activity *in situ*

Although the observed reduction in Rb^+ transport activity of the $\Delta13$ and $\Delta29$ truncated mutants appears moderate, it is important to note that these subtle effects were observed already at a relatively mild proton gradient ($pH_{ext}= 5.5$, $\Delta pH \sim 2$). However, *in situ* the H,K-ATPase pumps protons against a 10,000-fold higher gradient of about 10^6. Under these physiological conditions (which are unfortunately not applicable to *Xenopus* oocytes), the high lumenal H^+ concentration would even more favour H^+ reverse binding at the extracellular sites, thereby also stimulating the $E_2P \rightarrow E_1P$ "backward" reaction (i.e. increasing k_{rev}). An enhanced k_{rev} in addition to the reduced forward rate constant k_{fw} (as observed for the N-terminally deleted mutants $\Delta13$ and $\Delta29$) is thus expected to have even more drastic effects on the turnover number. Therefore, albeit causing only small effects at the rather shallow pH gradient applied here, the destabilization of the E_2-state by the N-terminal truncations will almost certainly slow down H^+ secretion under the pH conditions in the stomach.

5.3.2. Adverse conformational effects exerted by two contact sites between α-subunit and the β-N-terminus

A surprising finding from the VCF experiments is that the more extensive deletion β$\Delta29$ caused a smaller shift towards the E_1P-state than β$\Delta13$ (see $V_{0.5}$ values in Table 5.1 on page 72). The two putative interaction sites between the β-N-terminus and the α-subunit proposed by the recently published cryo-EM structure of the pig gastric H,K-ATPase may provide a rationale to explain these rather unexpected effects: Apart from the aforementioned contact between β-N-terminus and P-domain (probably around Arg-716 to Ala-719, see red arrow in Figures 1.5 and 5.5), a stretch of the β-N-terminus located closer to the transmembrane domain also approaches the cytoplasmic stalk of α-TM3, which is connected to the A-domain (Figures 1.5 and 5.5, black arrow). Therefore, a truncation of the first 12 amino acids in β$\Delta13$ may only disrupt the interaction with the P-domain and thereby cause the strong E_1P shift, whereas the more extensive deletion in β$\Delta29$ most likely affects both contacts, which might involve an additional compensatory mechanism, as follows.

As outlined by Abe and colleagues, the contact between the β-subunit N-terminus and the P-domain probably stabilizes the enzyme in E_2P, thereby minimizing the reverse reaction with ADP via $E_2P \rightarrow E_1P + ADP \rightarrow E_1 + ATP$. This was concluded from the putative changes in the arrangement of the N- and P-domain during the $E_1P \rightarrow E_2P$ conformational transition, as predicted by comparison between the cryo-EM structure of the gastric H,K-ATPase in the pseudo-E_2P conformation (Abe et al., 2009) and a SERCA structure corresponding to a E_1P-ADP intermediate (SERCA E_1AlF4-ADP structure by Toyoshima et al. (2004)). As illustrated in Figure 5.5, the phosphate analogue (AlF4) in the E_1P-ADP-like structure is in close proximity to the ADP molecule bound to the N-domain, whereas in the pseudo-E_2P structure of the H,K-ATPase the phosphate analogue is separated from the ADP molecule due to an inclination of the P-domain and an outward bending of the N-domain. Therefore, reverse transfer of the phosphate (bound to Asp-385 in the P-domain) to ADP (bound to the N-domain), would require a large movement of the P-domain, which is presumably prohibited by the tethering between P-domain and the β-N-terminus. This inter-subunit interaction of the H,K-ATPase was therefore interpreted in terms of a 'ratchet'-like mechanism (Abe et al., 2009), which is probably disrupted for the E_1-shifted, $\Delta13$ N-terminally deleted mutant.

Yet, another E_2-favoring effect might be exerted by the A-domain's TGES motif, which penetrates into the gap between the P- and the N-domain (shown in Figure 5.5, but also observed e.g. in the SERCA E_2P structure (Olesen et al., 2007)). Since the bound phosphate analog AlF4 in the P-domain is obscured by the protrusion of this TGES-motif (which segregates the N-domain far from the P-domain), neither water molecules nor ADP have access to the aspartylphosphate. This therefore prevents back-transfer of the phosphate to ADP. Truncation of 28 β-N-terminal amino acids (β$\Delta29$ mutant) probably also disrupts the aforementioned second contact between N-terminus and α-subunit (close to TM3, indicated by the black arrow in Figure), which may result in an unrestricted movement

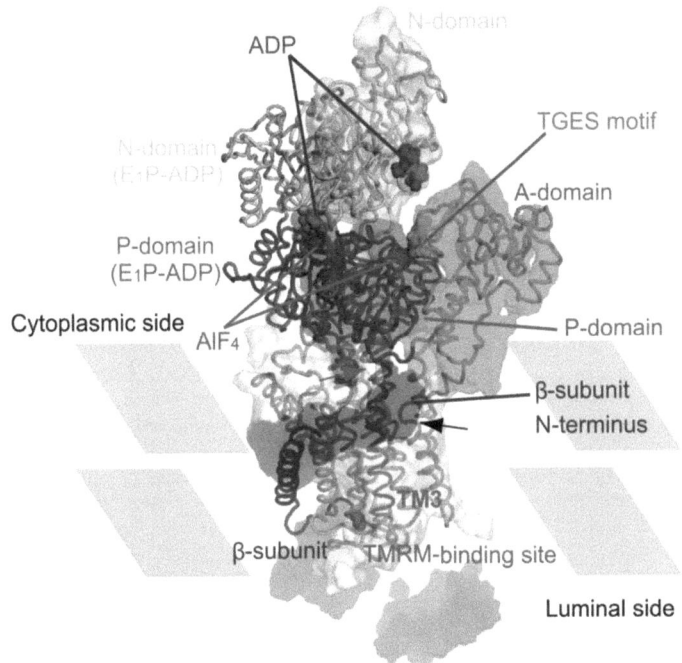

Figure 5.5.: **Potential changes in the arrangement of N- and P-domain during the $E_1P \rightarrow E_2P$ conformational transition of gastric H,K-ATPase.**[1] Comparison between the cryo-EM structure of the gastric H,K-ATPase in the AlF4-bound, pseudo-E_2P conformation (see also Figure 1.5) and the SERCA structure corresponding to an E_1P-ADP intermediate (SERCA E_1AlF4-ADP structure by Toyoshima et al. (2004), PDB code 2ZBD, showing the different orientation of N-domain and P-domain with respect to each other. Other parts of SERCA structure are omitted for clarity).

of the A-domain via the released TM3-A-linker. This in turn may lead to an enhanced segregating effect of the TGES protrusion of the A domain, thereby compensating partially for the E_1P-shift caused by the loss of the E_2-stabilizing contact between N-terminus and P-domain.

5.3.3. Comparison to the functional effects of β-N-terminally truncated Na,K-ATPase

In this chapter, for the first time the functional relevance of the β-subunits' N-terminal domain for H,K-ATPase transport activity was investigated *in vivo*. The results demonstrate that in analogy to β-N-terminally deleted Na,K-ATPase, N-terminal truncation of the H,K-ATPase β-subunit does not affect its chaperone-like functions, since the amount of α-subunit in the plasma membrane determined by Western blot analysis was virtually unchanged for all four investigated deletions. Thus, the N-teminus is apparently dispensable for α/β coassembly, maturation and plasma membrane targeting of both oligomeric ATPases.

As a key observation, the recently published cryo-EM structure of the gastric H,K-ATPase highlighted the close proximity between the P-domain of the α-subunit and the short N-terminal tail of the β-subunit, suggesting an E_2P-stabilizing interaction of the two subunits. The results from this chapter on N-terminally truncated β-variants provide direct evidence that the β-N-terminus indeed

---
[1]This figure was kindly provided by K. Abe, University of Kyoto, Japan.

construct	Boltzmann parameter curves		stationary pump currents
	$V_{0.5}$ (mV)	z_q	(nA)
Na,K$\alpha_1$wt/β_1S62C	-92.7 ± 7.7	0.70 ± 0.03	221 ± 85
Na,K$\alpha_1$wt/β_1S62CΔ8	-100.8 ± 7.5	0.68 ± 0.03	232 ± 38

Table 5.3.: **Stationary pump currents and parameters from fits of a Boltzmann function to (1-ΔF/F)-V distributions of wildtype or β-N-terminally deleted Na,K-ATPase.** Values are means ± S.D. of 10-15 oocytes.

assists in E_2P-state stabilization and that this is critical for the enzyme's transport efficiency under *in vivo* conditions.

Considering crystal structures of the closely related Na,K-ATPase in the E_2-state (Morth et al., 2007; Shinoda et al., 2009), there is no indication for similar interactions between β-N-terminus and α-subunit of the sodium pump. Since however the β-N-terminus of the Na,K-ATPase was not completely resolved in these structures, it does not really provide an answer to the question whether the inter-subunit interaction mediated by the H,K-ATPase β-N-terminus is conserved among P$_{2C}$-type ATPases.

As outlined in the introductory part of this chapter, truncation of almost the entire N-terminal part of the Na,K-ATPase β-subunit had profound effects on the transport properties of the enzyme: In a study from 1998, Hasler and colleagues demonstrated that the truncation reduced the apparent affinities for Na$^+$ and K$^+$, as determined by electrophysiological techniques (Hasler et al., 1998). Since however, shorter deletions and multiple point mutations in the N-terminus did not change the apparent ion affinities, the authors considered it more likely that the truncation rather influences the affinities indirectly by inducing conformational changes in other domains in the β-subunit. As several other studies showed that mainly the ectodomain is implicated in modulating the ion affinities, they concluded that the truncation must cause changes in this domain. Moreover, they proposed a shift in the position of the transmembrane domain within the membrane, because this was observed for N-terminal truncations in other type II proteins (asialoglycoprotein receptor and the invariant chain of MHCII antigen). As a matter of fact, they provided evidence for this hypothesis by "glycosylation mapping" two years later (Hasler et al., 2000). Of note, this extensive N-terminal truncation of the Na,K-ATPase β-subunit resulted in an E_2-shifted phenotype. According to these findings, a similar E_2-stabilizing effect of the Na,K-ATPase β-N-terminus does not seem very likely. However, potential conformational effects have not been determined yet for shorter N-terminal deletions of the Na,K-ATPase β-subunit.

Therefore, to complete the current investigation, we deleted the eight most N-terminal amino acids of reporter construct β_1S62C of the sheep Na,K-ATPase β-subunit. Since the cytoplasmic β-N-terminus of the Na,K-ATPase is shorter than that of the gastric H,K-ATPase, this truncation results in a N-terminal domain comparable in length to the $\beta\Delta13$ variant of the gastric H,K-ATPase (see 5.6 A). This N-terminally truncated β-variant NaKβ_1S62CΔ8 was coexpressed with the wildtype sheep α_1-subunit, labeled by TMRM and subjected to voltage jumps in a high Na$^+$/K$^+$-free solution. Notably, the resulting voltage-dependent distribution of fluorescence amplitudes is not significantly different from the one observed for full-length β_1S62C containing Na,K-ATPase complexes (see Figure 5.6 B and Table 5.3). Furthermore, as inferred from the stationary pump currents at 10 mM K$^+$, the ion transport activity of the Na,K-ATPase is completely unaffected by the β-N-terminal truncation (Figure 5.6 B, inset).

Regarding the comparably shallow Na$^+$ gradient under physiological conditions, an analogous 'ratchet' like E_2P-stabilization is probably not necessary to assist Na$^+$ transport of the sodium pump. This further supports our hypothesis that the E_2P-state stabilization mediated by the H,K-ATPase β-N-terminus is unique to facilitate proton transport against the steep proton gradient across the parietal cell membrane, which is approximately 10,000-fold higher than the typical transmembrane gradient for Na$^+$ ions. Therefore, the mechanism of E_2P-state stabilization by the gastric H,K-ATPase β-N-terminus appears to be a distinctive property of this enzyme.

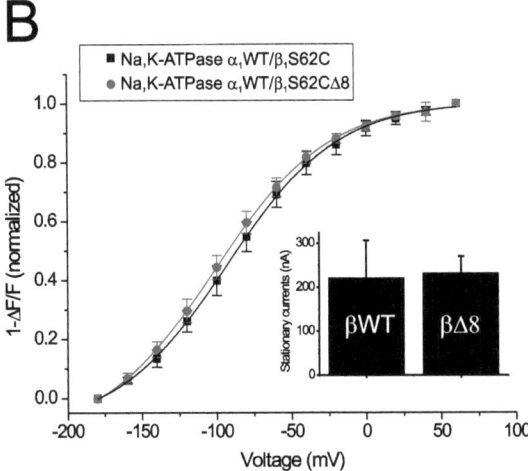

Figure 5.6.: **Conformational E_1P/E_2P distribution and pump currents of β-N-terminally truncated Na,K-ATPase.** A: Partial amino acid sequences of the N-terminally deleted Na,K-ATPase β-variant NaKβ$_1$Δ8 and the wildtype sheep Na,K-ATPase β$_1$-subunit. For comparison, the wildtype and N-terminally truncated β-variants of the rat gastric H,K-ATPase are also shown. B: Voltage-dependent distribution of fluorescence amplitudes 1-ΔF/F for Na,K-ATPase complexes consisting of the sheep wildtype α$_1$-subunit and either the unmodified reporter construct β$_1$S62C (squares), or the N-terminally deleted β-variant NaK-β$_1$Δ8 (circles). Inset: Stationary pump currents of the two constructs at saturating K^+ concentrations (10 mM). All data are means ± S.D. of 10-15 oocytes.

CHAPTER **6**

Summary and Conclusions

Summary

The Na,K-ATPase and gastric H,K-ATPase belong to the extensive class of P-type ATPases. These ATPases use ATP hydrolysis for active transport of cations. Gastric H,K-ATPase and the ubiquitous Na,K-ATPase share several similarities. The two ion pumps have closely related catalytic α-subunits which have a high sequence homology (approximately 60% sequence identity) and glycosylated β-subunits with lower sequence identity, but similar basic structural features. For example, the membrane topology is identical: all known Na,K-ATPase β-isoforms and the H,K-ATPase β-subunit are type II membrane proteins consisting of a short cytoplasmic N-terminus, a single transmembrane domain and a large extracellullar C-terminal domain with conserved disulfide bridges and N-glycosylation sites. The β-subunits of P_{2C}-type ATPases are essentially required for proper folding, maturation and plasma membrane targeting of the catalytic α-subunits. In addition, it was shown that the β-subunits influence the transport activity of the ion pumps, but the molecular mechanisms underlying this modulation are not well understood so far, especially for the less extensively studied gastric proton pump.

In order to gain better insight into which parts of the β-subunits are responsible for their influence on ion transport of the respective catalytic α-subunits, mutational changes were introduced in all three topogenic domains of the two β-subunits. Furthermore, since recent structural findings on the two ATPases have revealed potential interaction sites between α- and β-subunit, the functional consequences of mutations in the affected regions (including some in the α-subunit) were analyzed.

N-glycosylation of the ectodomain (Chapter 3)

In the large ectodomain, the highly conserved N-glycosylation sites were removed by Asn→Gln mutations in all available Asn-X-Ser/Thr sequence motifs to investigate whether the huge sugar moiety that contributes significantly to the molecular weight of the β-subunit influences the transport properties of the catalytic α-subunit. Notably, the loss of glycosylation did neither affect the β-subunits' chaperone-like functions nor the ion transport activity of Na,K- and H,K-ATPase complexes assembled from the glycosylation-deficient β-variants. Despite causing a dramatic loss in molecular weight, even more subtle enzymatic properties like the E_1P/E_2P conformational distribution or the kinetics of the $E_1P \rightarrow E_2P$ transition were not affected by the lack of N-glycosylation for both ATPases. This was especially surprising for the gastric proton pump, since previous studies which investigated glycosylation-deficient H,K β-variants in different expression systems showed severe effects on plasma membrane targeting, α/β coassembly and ATPase activity. Yet, a possible explanation for these discrepancies may be that the purification protocols used in these studies are responsible for the observed effects, since the lack of N-glycosylation may render the ATPase molecules more susceptible to inactivation during these procedures, whereas the almost native environment of intact *Xenopus* oocyte membranes may efficiently protect the enzyme. However, these observations also suggest that (at least for the gastric H,K-ATPase) the N-linked oligosaccharides may have a protective role *in vivo*, where body temperature, low lumenal pH and the presence of proteolytic enzymes could favour the inactivation of functional H,K-ATPase α/β complexes as well.

Conserved Tyrosines in the transmembrane domain (Chapter 4)

Inspired by a study of Hasler et al. (2001) which has shown that tryptophan replacements of two highly conserved tyrosines in the transmembrane domain of the Na,K-ATPase β-subunit had profound effects on the apparent ion affinites of the sodium pump, we examined whether this phenotype is linked to changes in the E_1P/E_2P conformational equilibrium. Indeed, the mutation β-(Y39W,Y43W) resulted in a substantial shift of the E_1P/E_2P distribution towards the E_1P-state, which could be attributed to an increased backward rate constant of the $E_1P \rightarrow E_2P$ conformational transition without affecting the forward rate constant. When the corresponding double tryptophan substitution β-(Y44W,Y48W) in the gastric H,K-ATPase β-transmembrane region was investigated, a similar shift towards E_1P was observed that was also reflected by the mutant's reduced sensitivity towards SCH 28080 in Rb^+ uptake measurements. Remarkably, for the gastric H,K-ATPase, the mutation had even more dramatic consequences for the apparent cation afffinities: the extracellular Rb^+ affinity was substantially reduced and H^+ release from the extracellular binding sites was severely affected, too.

Since the recently published Na,K-ATPase crystal structures of the Rb^+ occluded E_2-state revealed some residues in TM7 of the α-subunit which are at interaction distance to the two β-tyrosines (Morth et al., 2007; Shinoda et al., 2009), a potential E_2-specific contact to these amino acids could result in an enhanced stabilization of the E_2-state which is presumably disrupted by the bulky tryptophan replacements in the examined β-mutants. This idea was further substantiated by the finding that for both ATPases, shifts toward E_1P occured also for tryptophan replacements of some of these α-TM7 residues. Furthermore, the kinetic changes in transient currents or fluorescence changes observed for these Na,K- and H,K-ATPase α-TM7-mutants closely resembled those of the β-(Y39W, Y43W) mutant, thus indicating a common molecular origin for the E_1-shifted phenotypes of α-TM7- and β-TM-mutants.

Since all described effects were consistently found for both Na,K-ATPase and gastric H,K-ATPase variants, the proposed E_2P-stabilizing α/β interaction is probably conserved among P_{2C}-type ATPases, although it is apparently of higher functional relevance for the gastric H,K-ATPase (see conclusions).

Truncation of the cytosolic N-terminus (Chapter 5)

The research on N-terminally deleted β-variants was motivated by the recently published cryo-EM structure of the gastric H,K-ATPase in the pseudo E_2P-state (Abe et al., 2009). The electron density suggested a potential E_2P-stabilizing contact between β-N-terminus and the P-domain of the α-subunit that was corroborated by pulse-chase experiments. Since, however, these studies were carried out under rather unphysiological conditions (low ionic strength, 0°C), the functional relevance of this putative interaction was investigated in *Xenopus* oocytes. Using voltage-clamp fluorometry and SCH 28080 sensitivity measurements, we confirmed the E_1P-shifted phenotypes of two N-terminally deleted β-variants (HKβΔ13 and HKβΔ29). Furthermore, we showed that the forward rate constant of the $E_1P \rightarrow E_2P$ conformational transition is affected by the truncation, which results in a reduced turnover number of these two β-mutants.

Some unexpected quantitative differences in the E_1-shifts observed for the βΔ13 and βΔ29 mutants indicate that another putative contact between β-N-terminus and TM3-linker of the α-subunit revealed by the Cryo-EM structure may as well influence the E_1P/E_2P conformational equilibrium. Finally, since a similar N-terminal truncation of the Na,K-ATPase β-subunit did neither affect the dynamic E_1P/E_2P equilibrium nor the K^+ activated pump currents, the E_2P-stabilizing effect of the β-N-terminus is apparently a unique property of the gastric H,K-ATPase which may be linked to significant functional consequences (see conclusions).

Chapter 6. Summary and Conclusions

Conclusions

The research presented in this thesis has provided novel insights into the molecular interactions between α- and β-subunits of Na,K- and gastric H,K-ATPase, having important functional consequences for cation transport activity of the two related ATPases, especially for the gastric proton pump (as briefly outlined below).

It was demonstrated that contacts between regions in the catalytic α-subunit and the accessory β-subunit of P_{2C}-type ATPase exist that significantly contribute to a stabilization of the E_2P conformation. Destabilization of the E_2P-state caused by mutational changes in several parts of the oligomeric enzyme (β-TM, α-TM7 or the β-N-terminus, in case of the H,K-ATPase) affected either the apparent ion affinity or the turnover number. The E_1P-shifting mutations obviously reduce the dwell time in the E_2P-state which is in agreement with the mutants decreased sensitivity towards the E_2P-specific inhibitor SCH 28080. For the gastric H,K-ATPase, a stabilization of the E_2P-state might be critical for proton pumping for the following reason: proton exchange between the lumenally exposed proton-coordinating residues in the E_2P conformation and the stomach lumen is probably very slow, since their pK_a may be close to the extremely low lumenal pH under *in vivo* conditions. Therefore, a sufficiently long dwell time in the E_2P-state might be required to reach thermodynamic equilibrium. Since other P-Type ATPases are probably able to release their respective ions from the E_2P-state substantially faster due to the lower concentration gradients, molecular mechanisms that enhance a stabilization of this intermediate may be uniquely relevant for the gastric H,K-ATPase, although some E_2-stabilizing α/β interactions apparently also exist in the Na,K-ATPase.

Whereas these findings contributed to a better understanding of how the β-subunit of gastric H,K-ATPase may assist the catalytic α-subunit in fulfilling its highly demanding task of proton pumping against a million-fold proton gradient, the results from the studies on glycosylation-deficient Na,K- and H,K-ATPase β-mutants suggest an additional function of the H,K-ATPase β-subunit in protecting the protein from acidic denaturation or proteolytic digestion. Again, although N-glycosylation is conserved among Na,K- and H,K-ATPase β-subunits, a need for protective mechanisms exists only for the gastric H,K-ATPase, which operates in an extraordinarily harsh environment.

Acknowledgements

While I was working on this thesis, I enjoyed the help and company of many friendly individuals whom I would like to thank in the following section:

First of all, I gratefully acknowledge the excellent supervision of Prof. Dr. Thomas Friedrich. I am especially thankful for his encouragement, kindness, support, patience and valuable guidance throughout my thesis. Many thanks also to Prof. Dr. Ernst Bamberg, head of the department for Biophysical Chemistry at the Max-Planck-Institute for Biophysics in Frankfurt, where I started my Ph.D. work. Furthermore, I would like to thank these two people (and the Max-Planck-Gesellschaft zur Förderung der Wissenschaften, DFG SFB 472, SFB 740 and the Cluster of Excellence "UNICAT") for ensuring financial support throughout my Ph.D. work, for providing me with excellent facilities to pursue my experiments and for giving me many opportunities to present my research at various international conferences. I thank Prof. Dr. Peter Hildebrandt for reading this thesis and Prof. Prof. Dr. Roderich Süßmuth for presiding over my disputation.

I thank my friend and colleague Neslie, who also moved from Frankfurt to Berlin during her thesis, for sharing countless unforgettable experiences in the lab, the office and "elsevier". Special thanks to her for preparing numerous delicious "NesTeas". My research benefited enormously from our collaboration with Kazuhiro Abe from Kyoto University in Japan. His cryo-EM structure of the gastric H,K-ATPase was an invaluable inspiration for my investigations on the proton pump. Additionally, I would like to acknowledge all the beautiful graphics and the impressive artwork he prepared and kindly provided to us. I also greatly appreciate the numerous helpful suggestions by David von Stetten and Volker Mosthaf concerning all LaTeX typesetting issues which sometimes gave me a hard time when I was writing this thesis. I am grateful to all my other colleagues at the Max-Planck-Institute in Frankfurt and the Max-Volmer-Laboratory in Berlin for providing an excellent working atmosphere and inspiring discussions. Finally, I am forever indebted to my parents and Volker for their understanding, endless patience and encouragement.

CHAPTER **A**

Supplementary data

Sequence alignment of Na,K- and H,K-ATPase β-subunits

```
                                              10        20
NaKb_brine_shrimp    ....MADKKPD..EQFV..GSGPKET......KWQSFKGFVWNSETSQFM   36
NaKb_zebrafish       .............MSKK..SENQRRA.....KSREQLERRDLHPRTGELF   30
NaKb1_nematode       .....MEKKVDE.NATL..MNGVETTGPARDDVPETFREFLYNKKNGTVM   42
NaKb1_fruit_fly      .............MSKNNGKGAKGEFEFPQPAKKQTFSEMIYNPQEGTFF   37
NaKb1_electric_eel   ...............MARE..KSTDD........GGGWKKFLWDSEKKQVL   26
NaKb1_claw_frog      ...............MARD..KAKET........DGGWRKFIWNADKKEFL   26
NaKb1_chicken        ...............MARG..KANDG........DGNWKKFIWNSEKKELL   26
NaKb1_pig            ...............MARG..KAKE.........EGSWKKFIWNSEKKEFL   25
NaKb1_sheep          ...............MARG..KAKE.........EGSWKKFIWNSEKKEFL   25
NaKb1_rat            ...............MARG..KAKE.........EGSWKKFIWNSEKKEFL   25
NaKb1_human          ...............MARG..KAKE.........EGSWKKFIWNSEKKEFL   25
NaKb2_rat            ...............MVIQ..KEKKSCG.....QVVEEWKEFVWNPRTHQFM   30
NaKb2_human          ...............MVIQ..KEKKSCG.....QVVEEWKEFVWNPRTHQFM   30
NaKb3_claw_frog      ...............MAK...EENKGSE.....QSGSDWKQFIYNPQKGEFM   29
NaKb3_rat            ...............MTK...TEKKSFH.....QSLAEWKLFIYNPTSGEFL   29
NaKb3_human          ...............MTK...NEKKSLN.....QSLAEWKLFIYNPTTGEFL   29
HKb_claw_frog        ...............MATF..NEKKTCS.....QRMENFGRFVWNPDTSEFM   30
HKb_chicken          ...............MATL..NEKKTCS.....ERMENFRRFVWNPETKLFM   30
HKb_pig              ...............MAAL..QEKKSCS.....QRMEEFQRYCWNPDTGQML   30
HKb_rat              ...............MAAL..QEKKSCS.....QRMAEFRQYCWNPDTGQML   30
HKb_human            ...............MAAL..QEKKTCG.....QRMEEFQRYCWNPDTGQML   30

                                  transmembrane domain
                              30        40        50        60        70
NaKb_brine_shrimp    GRTAGSWAKITIFYVIFYTLLAGFFAGMLMIFYQTLDFK.IPKWQNKDSL   85
NaKb_zebrafish       GRTARNWGLILLFYLVFYGFLAAMFVFTLWVMLQTLNDD.TPKYRDR...   76
NaKb1_nematode       GRTGKSWFQIIVFYIIFYAFLAAWFLTCLTIFMKTLDPK.VPRFYGKGTI   91
NaKb1_fruit_fly      GRTGKSWSQLLLFYTIFYIVLAALFTICMQGLLSTISDT.EPKWKLQDSL   86
NaKb1_electric_eel   GRTGTSWFKIFVFYLIFYGCLAGIFIGTIQVMLLTISDF.EPKYQDR...   72
NaKb1_claw_frog      GRTGGSWFKILLFYLIFYGCLAGIFIGTIQVLLLTISEF.EPKYQDR...   72
NaKb1_chicken        GRTGGSWFKILLFYVIFYGCLAGIFIGTIQVMLLTVSEF.EPKYQDR...   72
NaKb1_pig            GRTGGSWFKILLFYVIFYGCLAGIFIGTIQVMLLTISEF.KPTYQDR...   71
NaKb1_sheep          GRTGGSWFKILLFYVIFYGCLAGIFIGTIQVMLLTISEF.KPTYQDR...   71
NaKb1_rat            GRTGGSWFKILLFYVIFYGCLAGIFIGTIQVMLLTISEL.KPTYQDR...   71
NaKb1_human          GRTGGSWFKILLFYVIFYGCLAGIFIGTIQVMLLTISEF.KPTYQDR...   71
NaKb2_rat            GRTGTSWAFILLFYLVFYGFLTAMFTLTMWVMLQTVSDH.TPKYQDR...   76
NaKb2_human          GRTGTSWAFILLFYLVFYGFPTAMFTLTMWVMLQTVSDH.TPKYQDR...   76
NaKb3_claw_frog      GRTASSWALILLFYLVFYGFLAGLFTLTMWVMLQTLDDS.VPKYRDR...   75
NaKb3_rat            GRTSKSWGLILLFYLVFYGFLAALFFTMWVMLQTLNDE.VPKYRDQ...   75
NaKb3_human          GRTAKSWGLILLFYLVFYGFLAALFSFTMWVMLQTLNDE.VPKYRDQ...   75
HKb_claw_frog        GRTFAKWVYISLYYAAFYVIMVGIFALSIYSLMQTLSPY.VPDYQDE...   76
HKb_chicken          GRTLINWVWISLYYLAFYVVMTGIFALSIYSLMRTVNPY.EPDYQDQ...   76
HKb_pig              GRTLSRWVWISLYYVAFYVVMSGIFALCIYVLMRTIDPY.TPDYQDQ...   76
HKb_rat              GRTPARWVWISLYYAAFYVVMTGLFALCIYVLMQTIDPY.TPDYQDQ...   76
HKb_human            GRTLSRWVWISLYYVAFYVVMTGLFALCLYVLMQTVDPY.TPDYQDQ...   76
                                 ▲      ▲
                                Tyr    Tyr
```

Sequence alignment of Na,K- and H,K-ATPase β-subunits

```
                        80         90        100       110
NaKb_brine_shrimp  IGANPGLGFRPMPPEAQVDSTLIQFKHG.IKGDWQYWVHSLTEFLEPYET  134
NaKb_zebrafish     VA.SPGLVIRPN.S......LNIEFNRS.DPLEYGQYVQHLESFLHQYND  117
NaKb1_nematode     IGVNPGVGYQPWLKER.PDSTLIKYNLR.DQKSYKAYLEQMKTYLTKYDS  139
NaKb1_fruit_fly    IGTNPGLGFRPLSEQTE.RGSVIAFDGK.KPAESDYWIELIDDFLRDYNH  134
NaKb1_electric_eel VA.PPGLSHSPYAV.....KTEISFSVS.NPNSYENHVNGLKELLKNYE   115
NaKb1_claw_frog    VA.PPGLTQLPRAV.....KTEISFSPS.DSNSYQEYVKSMDNFLSKYNN  115
NaKb1_chicken      VA.PPGLTQVPQVQ.....KTEISFTVN.DPKSYDPYVKNLEGFLNKYSA  115
NaKb1_pig          VA.PPGLTQIPQSQ.....KTEISFRPN.DPQSYESYVVSIVRFLEKYKD  114
NaKb1_sheep        VA.PPGLTQIPQIQ.....KTEIAFRPN.DPKSYMTYVDNIDNFLKKYRD  114
NaKb1_rat          VA.PPGLTQIPQIQ.....KTEISFRPN.DPKSYEAYVLNIIRFLEKYKD  114
NaKb1_human        VA.PPGLTQIPQIQ.....KTEISFRPN.DPKSYEAYVLNIVRFLEKYKD  114
NaKb2_rat          LA.TPGLMIRPKTEN.....LDVIVNIS.DTESWDQHVQKLNKFLEPYND  119
NaKb2_human        LA.TPGLMIRPKTEN.....LDVIVNVS.DTESWDQHVQKLNKFLEPYND  119
NaKb3_claw_frog    VS.SPGLMISPK.S......AGLEIKFSRS.KTQSYMEYVQTLNTFLAPYND 118
NaKb3_rat          IP.SPGLMVFPKPP....TALDYTYSMS.DPHTYKKFVEDLKNFLKPYSV  119
NaKb3_human        IP.SPGLMVFPKPV....TALEYTFSRS.DPTSYAGYIEDLKKFLKPYTL  119
HKb_claw_frog      LK.SPGVTLRPDPYGD..EVIELFYNMA.DNKTYLPLVTSLCEFLPVYNK  122
HKb_chicken        LK.SPGVTLRPDVYGH..RGLQIYYNAS.DNKTWEGLVTMLQTFLTAYSP  122
HKb_pig            LK.SPGVTLRPDVYGE..KGLDISYNVS.DSTTWAGLAHTLHRFLAGYSP  122
HKb_rat            LK.SPGVTLRPDVYGE..RGLQISYNIS.ENSSWAGLTHTLHSFLAGYTP  122
HKb_human          LR.SPGVTLRPDVYGE..KGLEIVYNVS.DNRTWADLTQTLHAFLAGYSP  122

                       120       130       140       150
NaKb_brine_shrimp  LTSS...GQEFTNCDFDKP..........PQEGKACNFNVELLGDH...  167
NaKb_zebrafish     SEQAK.....NDLCYGGTVPEQDG......ESLKKVCQFKRSLLYS....  152
NaKb1_nematode     NATETRE......CGAGDSNDDLEK.....NPDALPCRFDLSVFDKG...  175
NaKb1_fruit_fly    TEGRD.....MKHCGFGQVLEPT..........DVCVVNTDLFGG....  164
NaKb1_electric_eel SKQDG..NTPFEDCGVIPADYITRGPIEESQGGKRVCRFLLQWLKN....  159
NaKb1_claw_frog    EKQ.G..SNMFEDCGTIPGPYHERGALNKDEGMKKSCVFRREWLGN....  158
NaKb1_chicken      GEQTD..NIVFQDCGDIPTDYKERGPYNDAQGGQKKVCKFKREWLEN...  159
NaKb1_pig          LAQKD..DMIFEDCGNVPSELKERGEYNNERGERKVCRFRLEWLGN....  158
NaKb1_sheep        SAQKD..DMIFEDCGNVPSELKDRGEFNNEQGERKVCRFKLEWLGN....  158
NaKb1_rat          SAQKD..DMIFEDCGSMPSEPKERGEFNHERGERKVCRFKLDWLGN....  158
NaKb1_human        SAQRD..DMIFEDCGDVPSEPKERGDFNHERGERKVCRFKLEWLGN....  158
NaKb2_rat          SIQAQ....KNDVCRPGRYYEQPDNG..VLNYPKRACQFNRTQLGN....  159
NaKb2_human        SMQAQ....KNDVCRPGRYYEQPDNG..VLNYPKLACQFNRTQLGN....  159
NaKb3_claw_frog    SIQAK.....NEFCPPPGLYFDQDEE......VEKKTCQFNRTSLGI....  153
NaKb3_rat          EEQKN.....LTDCPGGALFHQE.......GPDYSACQFPVSLLQE....  153
NaKb3_human        EEQKN.....LTVCPDGALFEQK.......GPVYVACQFPISLLQA....  153
HKb_claw_frog      SVQEK.....MNANCSDHTRISCAHQ.....NENTKSCQFTTDMLGN....  159
HKb_chicken        AAQHL.....NINCTSNTYFIQNTFD..GPNNTKLSCKFTSDMLQN....  161
HKb_pig            AAQEG.....SINCTSEKYFFQESFL..APNHTKFSCKFTADMLQN....  161
HKb_rat            ASQQD.....SINCSSEKYFFQETFS..APNHTKFSCKFTADMLQN....  161
HKb_human          AAQED.....SINCTSEQYFFQESFR..APNHTKFSCKFTADMLQN....  161
                               ▲                    ▲
                             Cys1                 Cys2
```

```
                            160       170       180            190
NaKb_brine_shrimp    ..CTKENN...FGYELGKPCVLIKLTDFG.......WRPEVY.NSSAEVP  204
NaKb_zebrafish       ..CSGMEDT.TFGYAKGQPCVIVKMNRIIG......LKPS..GD......  185
NaKb1_nematode       ..CSEKSD...FGYKSGKPCVIISLNRLIG......WRPTDY.QE.NSVP  212
NaKb1_fruit_fly      ..CSKANN...YGYKTNQPCIFLKLNKIFG......WIPEVYDKEEKDMP  203
NaKb1_electric_eel   ..CSGIDD.PSYGYSEGKPCIIAKLNRILG......FYPKPPKNGTDLPE  200
NaKb1_claw_frog      ..CSGLND.PSYGFADGKPCVIVKLNRILA......FKPVPPQNNS.LPP  198
NaKb1_chicken        ..CSGLQD.NTFGYKDGKPCILVKLNRIIG......FKPKAPENESLPSD  200
NaKb1_pig            ..CSGLND.ETYGYKDGKPCVIIKLNRVLG......FKPKPPKNESLETY  199
NaKb1_sheep          ..CSGIND.ETYGYKEGKPCVIIKLNRVLG......FKPKPPKNESLETY  199
NaKb1_rat            ..CSGLND.ESYGYKEGKPCIIIKLNRVLG......FKPKPPKNESLETY  199
NaKb1_human          ..CSGLND.ETYGYKEGKPCIIIKLNRVLG......FKPKPPKNESLETY  199
NaKb2_rat            ..CSGIGDPTHYGYSTGQPCVFIKMNRVIN......FYAG..ANQS....  195
NaKb2_human          ..CSGIGDSTHYGYSTGQPCVFIKMNRVIN......FYAG..ANQS....  195
NaKb3_claw_frog      ..CSGIEDP.MFGYGEGKPCVIVKINRIIG......LKPE..GN......  186
NaKb3_rat            ..CSGVNDS.NFGYSKGQPCVLVKMNRIIE......LVPD..GA......  186
NaKb3_human          ..CSGMNDP.DFGYSQGNPCILVKMNRIIG......LKPE..GV......  186
HKb_claw_frog        ..CSWEHD.HTFGYKSGKPCLFIKMNRIIN......FVPG...NKTA...  194
HKb_chicken          ..CSGITD.PTFGFPEGKPCFIVKMNRIIK......FYPG...NGTA...  196
HKb_pig              ..CSGRPD.PTFGFAEGKPCFIIKMNRIVK......FLPG...NSTA...  196
HKb_rat              ..CSGLVD.PSFGFEEGKPCFIIKMNRIVK......FLPS...NNTA...  196
HKb_human            ..CSGLAD.PNFGFEEGKPCFIIKMNRIVK......FLPS...NGSA...  196
                         ▲                  ▲
                       Cys3               Cys4

                        200       210       220       230
NaKb_brine_shrimp    EDMPADLKSYIKDIETGNKTHMNMVWLSCEGETAND....KEKIGTITYT  250
NaKb_zebrafish       ....................PYINCTSKSVK.......PLQMTSI     203
NaKb1_nematode       EEVKDRYKAGS............IAINCRGATNVD....QEHIGKVTYM  245
NaKb1_fruit_fly      DDLKKVINETKTEER.......HEVWVSCFGHLGKD....KENFQNIRYF  242
NaKb1_electric_eel   ALQANYNQYV.............LPIHCQAKKEED....KVRIGTIEYF  232
NaKb1_claw_frog      EMTLNYNPYV.............IPIHCQAKKEED....IEKIKEVKYY  230
NaKb1_chicken        L.AGKYNPYL.............IPVHCVAKRDED....ADKIGMVEYY  231
NaKb1_pig            P.VMKYNPYV.............LPVHCTGKRDED....KEKVGTMEYF  230
NaKb1_sheep          P.VMKYNPYV.............LPVQCTGKRDED....KEKVGSIEYF  230
NaKb1_rat            PLTMKYNPNV.............LPVQCTGKRDED....KDKVGNIEYF  231
NaKb1_human          P.VMKYNPNV.............LPVQCTGKRDED....KDKVGNVEYF  230
NaKb2_rat            .......................MNVTCVGKKDED....AENLGHFIMF  217
NaKb2_human          .......................MNVTCAGKRDED....AENLGHFIMF  217
NaKb3_claw_frog      ......................PKINCTSKTE.......DVNLQYF     203
NaKb3_rat            ......................PYITCITKEEN........IANIVTY  204
NaKb3_human          ......................PRIDCVSKNED........IPNVAVY  204
HKb_claw_frog        ......................PLVNCSAENGE........LGDVQYY  212
HKb_chicken          ......................PRVDCSYVGDE......SRPLEVEYY  216
HKb_pig              ......................PRVDCAFLDQPR....DGPPLQVEYF  218
HKb_rat              ......................PRVDCTFQDDPQKPRKDIEPLQVQYY  222
HKb_human            ......................PRVDCAFLDQPRE...LGQPLQVKYY  219
                                              ▲
                                            Cys5
```

Sequence alignment of Na,K- and H,K-ATPase β-subunits

```
                         240        250        260        270
NaKb_brine_shrimp   PFR......GFPAYYYPYL...NVPGYLTPVVALQFG..SLQNGQAVNVE  289
NaKb_zebrafish      SPM.....RTIDRMYFPYYGKKTHKGYVQPLVAVKLLLKKEDYNSELIVE  248
NaKb1_nematode      PSN......GIDGRYYPYV...FTKGYQQPIAMVKFD..TIPRNKLVIVE  284
NaKb1_fruit_fly     PSQ......GFPSYYYPFLN...QPGYLSPLVAVQFN..SPPKGQMLDVE  281
NaKb1_electric_eel  GMGG...VGGFPLQYYPYYGKRLQKNYLQPLVGIQFT..NLTHNVELRVE  277
NaKb1_claw_frog     GMGG...FAGFPLTYYPYYGKLLQPDYLQPLIAVQFT..NITFDAEVRIE  275
NaKb1_chicken       GMGG...YPGFALQYYPYYGRLLQPQYLQPLVAVQFT..NLTYDVEVRVE  276
NaKb1_pig           GLGG...YPGFPLQYYPYYGKLLQPKYLQPLMAVQFT..NLTMDTEIRIE  275
NaKb1_sheep         GLGG...YPGFPLQYYPYYGKLLQPKYLQPLLAVQFT..NLTMDTEIRIE  275
NaKb1_rat           GMGG...FYGFPLQYYPYYGKLLQPKYLQPLLAVQFT..NLTLDTEIRIE  276
NaKb1_human         GLGN...SPGFPLQYYPYYGKLLQPKYLQPLLAVQFT..NLTMDTEIRIE  275
NaKb2_rat           PAN.....GNIDLMYFPYYGKKFHVNYTQPLVAVKFL..NVTPNVEVNVE  260
NaKb2_human         PAN.....GNIDLMYFPYYGKKFHVNYTQPLVAVKFL..NVTPNVEVNVE  260
NaKb3_claw_frog     PDN.....GKIDLMYFPYYGKKTHVNYVQPVVAVKISPSN.FTSEEIAVE  247
NaKb3_rat           PDD.....GLIDLKYFPYYGKKRHVGYRQPLVAVQVIFGADATKKEVTIE  249
NaKb3_human         PHN.....GMIDLKYFPYYGKKTHVGYQQPLVAVQVSFAPNNTGKEVTVE  249
HKb_claw_frog       PGND..TYGTIGLQYFPYCGKKMQPNYTNPLVAVKLL..NPTLNKELSVV  258
HKb_chicken         PVN.....GTFNLHYFPYYGKKAQPTYSNPLVAVKFL..NITKNVEVQIV  259
HKb_pig             PAN.....GTYSLHYFPYYGKKAQPHYSNPLVAAKLL..NVPRNRDVVIV  261
HKb_rat             PPN.....GTFSLHYFPYYGKKAQPHYSNPLVAAKFL..NVPKNTQVLIV  265
HKb_human           PPN.....GTFSLHYFPYYGKKAQPHYSNPLVAAKLL..NIPRNAEVAIV  262

                      280         290        300
NaKb_brine_shrimp   CKAWAN.NISRDR..QRRLGSVHFEIRMD............  315
NaKb_zebrafish      CKVEGS.NLKNNDERDKFLGRVTFRVLVTE............  277
NaKb1_nematode      CRAYAL.NIEHDI..SSRLGMVYFEVMVEDK.PVEEKKEL..  320
NaKb1_fruit_fly     CRAWAK.NIQYSGSARDRKGSVTFQILLD.............  309
NaKb1_electric_eel  CKVFGD.NIAYSE.KDRSLGRFEVKIEVKS............  305
NaKb1_claw_frog     CKAYGE.NIDYHD.KDRFQGRFDVKFDIKSS...........  304
NaKb1_chicken       CKEYGQ.NIQYSD.KDRFQGRFDIKFDIKSS...........  305
NaKb1_pig           CKAYGE.NIGYSE.KDRFQGRFDVKIEVKS............  303
NaKb1_sheep         CKAYGE.NIGYSE.KDRFQGRFDVKIEVKS............  303
NaKb1_rat           CKAYGE.NIGYSE.KDRFQGRFDVKIEVKS............  304
NaKb1_human         CKAYGE.NIGYSE.KDRFQGRFDVKIEVKS............  303
NaKb2_rat           CRINAA.NIATDDERDKFAARVAFKLRINKA...........  290
NaKb2_human         CRINAA.NIATDDERDKFAGRVAFKLRINKT...........  290
NaKb3_claw_frog     CKIHGSRNLKNEDERDKFLGRVTFKVKITE............  277
NaKb3_rat           CQIDGTRNLKNKNERDKFLGRVSFKVIAHA............  279
NaKb3_human         CKIDGSANLKSQDDRDKFLGRVMFKITARA............  279
HKb_claw_frog       CKVSGH.GITSDNPHDPYEGKVSFKLKIENKPLSSSAN...   295
HKb_chicken         CKIIGA.GITFDNVHDPYEGKVEFKLKIEDGAARDSTKKHV   299
HKb_pig             CKILAE.HVSFDNPHDPYEGKVEFKLKIQK............  290
HKb_rat             CKIMAD.HVTFDNPHDPYEGKVEFKLTIQK............  294
HKb_human           CKVMAE.HVTFNNPHDPYEGKVEFKLKIEK............  291
                    ▲
                    Cys6
```

☒ non conserved
X ≥ 75% conserved
X all match

Figure A.1: **Sequence alignment of several Na,K-ATPase β-isoforms and H,K-ATPase β-subunits from different species.** The sequence alignment was adjusted manually according to Axelsen & Palmgren (1998). Top numbering refers to the sheep Na,K-ATPase β_1-isoform. N-Glycosylation sites are shaded in orange, conserved regions in green as indicated in the legend. Black triangles show the location of the 6 extracellular cysteines that are linked in disulfide bridges. The two highly conserved tyrosines in the transmembrane domain are enclosed in red frames. The three residues that are suitable for environmentally sensitive TMRM-labeling upon mutation to cysteines are shaded in magenta in the sheep Na,K-ATPase β_1-isoform sequence. The figure was prepared in LaTeX with TeXshade (Beitz, 2000).

List of Figures

1.1.	Post-Albers reaction cycle of the Na,K-ATPase	8
1.2.	Electrogenic steps in the H,K-ATPase and Na,K-ATPase reaction cycles	9
1.3.	Crystal structure of the pig renal Na,K-ATPase	11
1.4.	Na,K-ATPase crystal structure from shark rectal glands	12
1.5.	Structural representation of the pig gastric H,K-ATPase based on the recently published cryo-EM structure	13
1.6.	Chemical structures of cardiac glycosides	18
1.7.	A close-up view of the low-affinity ouabain binding site of Na,K-ATPase from shark rectal glands	20
1.8.	Schematic illustration of a gland in the gastric mucosa layer of the stomach	21
1.9.	The mechanism of activation of PPIs	25
1.10.	Structure and proposed inhibition mechanism for APAs	27
2.1.	Rb^+ transport properties of the H,K-ATPase wildtype and reporter construct HKαS806C	32
3.1.	N-glycosylation sites and cysteine mutations for site-specific labeling of Na,K- and H,K-ATPase	38
3.2.	Western blot analysis of plasma membrane and total membrane fractions from Na,K- and H,K-ATPase-expressing oocytes	39
3.3.	Rb^+ uptake into oocytes by wildtype and glycosylation-deficient H,K-ATPase enzymes	42
3.4.	Voltage-pulse induced fluorescence changes of TMRM-labeled Na,K-ATPase	43
3.5.	Voltage dependence of E_1P/E_2P distribution and kinetics of transitions for glycosylated or nonglycosylated Na,K-ATPase enzymes	44
3.6.	E_1P/E_2P distribution and kinetics of the E_1P/E_2P transition for wildtype or nonglycosylated H,K-ATPase complexes	45
4.1.	Structural representation of the transmembrane α/β interface of pig renal Na,K-ATPase and alignments of several Na,K- and H,K-ATPase α-TM7 and β-TM	51
4.2.	Voltage-pulse induced fluorescence changes and transient currents of site-specifically labeled Na,K-ATPase wildtype and (Y39W,Y43W) mutant enzymes	53
4.3.	Voltage dependence of the E_1P/E_2P distribution and kinetics of the $E_1P\rightarrow E_2P$ transition for Na,K-ATPase wildtype and (Y39W,Y43W) β-variant enzymes	54
4.4.	Determination of the apparent $K_{0.5}$ for extracellular K^+ of Na,K-ATPase $\alpha_1$wt/β_1S62C and $\alpha_1$wt/β_1S62C(Y39W,Y43W) variant enzymes by fluorescence titration experiments	55
4.5.	Functional properties of the gastric H,K-ATPase Y44W,Y48W mutant enzyme	57
4.6.	Acidification assay for gastric H,K-ATPase-expressing oocytes	58
4.7.	Schematic illustrations of the putative changes in the E_1P/E_2P conformational and external ion binding equilibria caused by the tryptophan replacements in the Na,K-ATPase and H,K-ATPase β-TM	60
4.8.	Voltage-dependent E_1P/E_2P distribution of Na,K- and H,K-ATPase α-TM7 variants	62
4.9.	Apparent affinity for extracellular K^+ of the Na,K-ATPase α_1-TM7 mutant Y847W	63
4.10.	Plasma membrane delivery of the H,K-ATPase β-(Y44W,Y48W) mutant	65

List of Figures

5.1. Partial amino acid sequence of the wildtype and several N-terminally truncated mutant β-subunits of rat gastric H,K-ATPase 69
5.2. Western Blot analysis of N-terminally-deleted H,K-ATPase β-subunit variants 70
5.3. Voltage-dependent E_1P/E_2P distribution and kinetics of the $E_1P \rightarrow E_2P$ transition of N-terminally truncated H,K-ATPase β-variants 71
5.4. Rb^+ uptake, SCH 28080 sensitivity and apparent Rb^+ affinity of β-N-terminally truncated H,K-variants 73
5.5. Potential changes in the arrangement of the N- and P-domain during the $E_1P \rightarrow E_2P$ conformational transition of gastric H,K-ATPase 75
5.6. Conformational E_1P/E_2P distribution and pump currents of β-N-terminally truncated Na,K-ATPase 77

A.1. Sequence alignment of several Na,K-ATPase β-isoforms and H,K-ATPase β-subunits from different species 89

List of Tables

1.1. Classification of P-Type ATPases 6
1.2. Tissue-specific expression of different Na,K-ATPase α-, β- and γ-isoforms 16
1.3. Different Proton Pump Inhibitors and their sites of reaction on the H,K-ATPase α-subunit 26

2.1. Temperature protocol for THGF-AAS 35

3.1. Stationary currents and turnover number of Na,K-ATPase α/β complexes containing wildtype or non-glycosylated β-subunits 41
3.2. Boltzmann Parameters for Q-V distributions of Na,K-ATPase and for $(1-\Delta F/F)$-V distributions of Na,K-ATPase or H,K-ATPase enzymes with wildtype or glycosylation-deficient β-subunits 46

4.1. Boltzmann parameters for Q-V distributions of Na,K-ATPase and $(1-\Delta F/F)$-V distributions of Na,K-ATPase or H,K-ATPase β-TM-variants 56
4.2. Boltzmann parameters for Q-V distributions of Na,K-ATPase and $(1-\Delta F/F)$-V distributions of Na,K-ATPase or H,K-ATPase α-TM7-variants 63
4.3. Apparent $K_{0.5}$ values for halfmaximal Rb^+ activation of H,K-ATPase α-TM7-variants 64

5.1. Parameters from fits of a Boltzmann function to $(1-\Delta F/F)$-V distributions of N-terminally deleted H,K-ATPase β-mutants 72
5.2. Normalized values for Rb^+ uptake, SCH 28080 sensitivity and apparent Rb^+ affinity of N-terminally truncated H,K-ATPase β-variants 73
5.3. Stationary pump currents and parameters from fits of a Boltzmann function to $(1-\Delta F/F)$-V distributions of wildtype or β-N-terminally deleted Na,K-ATPase 76

Bibliography

Abe, K., K. Tani, T. Nishizawa and Y. Fujiyoshi. 2009. "Inter-subunit interaction of gastric H^+,K^+-ATPase prevents reverse reaction of the transport cycle." *EMBO J.* 28:1637–1643.

Abe, K., S. Kaya, K Taniguchi, Y. Hayashi, T. Imagawa, M. Kikumoto, K. Oiwa and K. Sakaguchi. 2005. "Evidence for a relationship between activity and the tetraprotomeric assembly of solubilized pig gastric H^+,K^+-ATPase." *J. Biochem.* 138:293–301.

Abe, K., S. Kaya, T. Imagawa and K. Taniguchi. 2002. "Gastric H^+,K^+-ATPase liberates two moles of P_i from one mole of phosphoenzyme formed from a high-affinity ATP binding site and one mole of enzyme-bound ATP at the low-affinity site during cross-talk between catalytic subunits." *Biochemistry* 41:2438–2445.

Abe, K., S. Kaya, Y. Hayashi, T. Imagawa, M. Kikumoto, K. Oiwa, T. Katoh, M. Yazawa and K. Taniguchi. 2003. "Correlation between the activities and the oligomeric forms of pig gastric H^+,K^+-ATPase." *Biochemistry* 42:15132–15138.

Abriel, H., U. Hasler, K. Geering and J. D. Horisberger. 1999. "Role of the intracellular domain of the β-subunit in Na^+,K^+ pump function." *Biochim. Biophys. Acta* 1418:85–96.

Ackermann, U. and K. Geering. 1990. "Mutual dependence of Na^+,K^+-ATPase α- and β-subunits for correct posttranslational processing and intracellular transport." *FEBS Lett.* 269:105–108.

Albers, R. W. 1967. "Biochemical aspects of active transport." *Annu. Rev. Biochem.* 36:727–756.

Asano, S., K. Kawada, T. Kimura, A. V. Grishin, M. J. Caplan and N. Takeguchi. 2000. "The roles of carbohydrate chains of the β-subunit on the functional expression of gastric H^+,K^+-ATPase." *J. Biol. Chem.* 275:8324–8330.

Asano, S., T. Kimura, S. Ueno, M. Kawamura and N. Takeguchi. 1999. "Chimeric domain analysis of the compatibility between H^+, K^+-ATPase and Na^+,K^+-ATPase β-subunits for the functional expression of gastric H^+,K^+-ATPase." *J. Biol. Chem.* 274:22257–22265.

Askari, A. and W. Huang. 1980. "Na^+,K^+-ATPase: half-of-the-subunits cross-linking reactivity suggests an oligomeric structure containing a minimum of four catalytic subunits." *Biochem. Biophys. Res. Commun.* 93:448–453.

Axelsen, K. B. and M. G. Palmgren. 1998. "Evolution of substrate specificities in the P-type ATPase superfamily." *J. Mol. Evol.* 46:84–101.

Ball, W. J., Jr., A. Abbott, Y. Sun and B. Malik. 1992. "Monoclonal antibodies and the identification of functional regions of the Na^+,K^+-ATPase." *Ann. N. Y. Acad. Sci.* 671:436–9.

Barwe, S. P., S. Kim, S. A. Rajasekaran, J. U. Bowie and A. K. Rajasekaran. 2007. "Janus model of the Na^+,K^+-ATPase β-subunit transmembrane domain: distinct faces mediate α/β assembly and β-β homo-oligomerization." *J. Mol. Biol.* 365:706–714.

Beesley, R. C. and J. G. Forte. 1973. "Glycoproteins and glycolipids of oxyntic cell microsomes. I. Glycoproteins: carbohydrate composition, analytical and preparative fractionation." *Biochim Biophys Acta* 307:372–85.

Beggah, A. T., P. Jaunin and K. Geering. 1997. "Role of glycosylation and disulfide bond formation in the β-subunit in the folding and functional expression of Na^+,K^+-ATPase." *J. Biol. Chem.* 272:10318–10326.

Beguin, P., X. Wang, D. Firsov, A. Puoti, D. Claeys, J. D. Horisberger and K. Geering. 1997. "The γ -subunit is a specific component of the Na^+,K^+-ATPase and modulates its transport function." *EMBO J.* 16:4250–60.

Beil, W., I. Hackbarth and K. F. Sewing. 1986. "Mechanism of gastric antisecretory effect of SCH 28080." *Br. J. Pharmacol.* 88:19–23.

Beitz, E. 2000. "TEXshade: shading and labeling of multiple sequence alignments using LaTeX2 epsilon." *Bioinformatics* 16:135–139.

Besancon, M., A. Simon, G. Sachs and J. M. Shin. 1997. "Sites of reaction of the gastric H^+,K^+-ATPase with extracytoplasmic thiol reagents." *J. Biol. Chem.* 272:22438–22446.

Besancon, M., J. M. Shin, F. Mercier, K. Munson, M. Miller, S. Hersey and G. Sachs. 1993. "Membrane topology and omeprazole labeling of the gastric H^+,K^+-adenosinetriphosphatase." *Biochemistry* 32:2345–2355.

Bibert, S., S. Roy, D. Schaer, J. D. Horisberger and K. Geering. 2008. "Phosphorylation of phospholemman (FXYD1) by protein kinases A and C modulates distinct Na^+,K^+-ATPase isozymes." *J. Biol. Chem.* 283:476–486.

Black, J. W., W. A. Duncan, C. J. Durant, C. R. Ganellin and E. M. Parsons. 1972. "Definition and antagonism of histamine H_2 -receptors." *Nature* 236:385–390.

Blanco-Arias, P., A. P. Einholm, H. Mamsa, C. Concheiro, H. Gutiérrez-de Terán, J. Romero, M. S. Toustrup-Jensen, A. Carracedo, J. C Jen, B. Vilsen and M. J. Sobrido. 2009. "A C-terminal mutation of ATP1A3 underscores the crucial role of sodium affinity in the pathophysiology of rapid-onset dystonia-parkinsonism." *Hum. Mol. Genet.* 18:2370–2377.

Blanco, G., A. W. DeTomaso, J. Koster, Z. J. Xie and R. W. Mercer. 1994a. "The α-subunit of the Na^+,K^+-ATPase has catalytic activity independent of the β-subunit." *J. Biol. Chem.* 269:23420–23425.

Blanco, G., J. C. Koster and R. W. Mercer. 1994b. "The α-subunit of the Na,K-ATPase specifically and stably associates into oligomers." *Proc. Natl. Acad. Sci. U. S. A.* 91:8542–8546.

Blanco, G. and R. W. Mercer. 1998. "Isozymes of the Na^+-K^+-ATPase: heterogeneity in structure, diversity in function." *Am. J. Physiol.* 275:F633–F650.

Brotherus, J. R., L. Jacobsen and P. L. Jorgensen. 1983. "Soluble and enzymatically stable Na^+/K^+-ATPase from mammalian kidney consisting predominantly of protomer α/ β-units. Preparation, assay and reconstitution of active Na^+, K^+ transport." *Biochim. Biophys. Acta.* 731:290–303.

Buffin-Meyer, B., M. Younes-Ibrahim, C. Barlet-Bas, L. Cheval, S. Marsy and A. Doucet. 1997. "K $^+$depletion modifies the properties of SCH 28080-sensitive K^+-ATPase in rat collecting duct." *Am. J. Physiol.* 272:F124–F131.

Bull, P. C., G. R. Thomas, J. M. Rommens, J. R. Forbes and D. W. Cox. 1993. "The Wilson disease gene is a putative copper transporting P-type ATPase similar to the Menkes gene." *Nat. Genet.* 5:327–337.

Burley, S. K. and G. A. Petsko. 1985. "Aromatic-aromatic interaction: a mechanism of protein structure stabilization." *Science* 229:23–28.

Burley, S. K. and G. A. Petsko. 1986. "Amino-aromatic interactions in proteins." *FEBS Lett.* 203:139–143.

Burnay, M., G. Crambert, S. Kharoubi-Hess, K. Geering and J. D. Horisberger. 2001. "*Bufo marinus* bladder H^+/K^+-ATPase carries out electroneutral ion transport." *Am. J. Physiol. Renal. Physiol.* 281:F869–F874.

Buxbaum, E. and W. Schoner. 1991. "Phosphate binding and ATP-binding sites coexist in Na^+,K^+-transporting ATPase, as demonstrated by the inactivating $MgPO_4$ complex analogue $Co(NH_3)_4 PO_4$." *Eur. J. Biochem.* 195:407–419.

Cantilina, T., Y. Sagara, G. Inesi and L. R. Jones. 1993. "Comparative studies of cardiac and skeletal sarcoplasmic reticulum ATPases. Effect of a phospholamban antibody on enzyme activation by Ca^{2+}." *J. Biol. Chem.* 268:17018–17025.

Chow, D. C., C. M. Browning and J. G. Forte. 1992. "Gastric H^+,K^+-ATPase activity is inhibited by reduction of disulfide bonds in β-subunit." *Am. J. Physiol.* 263:C39–C46.

Chow, D. C. and J. G. Forte. 1993. "Characterization of the β-subunit of the H^+-K^+-ATPase using an inhibitory monoclonal antibody." *Am. J. Physiol.* 265:C1562–C1570.

Chow, D. C. and J. G. Forte. 1995. "Functional significance of the β-subunit for heterodimeric P-type ATPases." *J. Exp. Biol.* 198:1–17.

Clarke, R. J and D.J. Kane. 2007. "Two gears of pumping by the sodium pump." *Biophys. J.* 93:4187–4196.

Codina, J., T. A. Pressley and T. D. DuBose. 1999. "The colonic H^+,K^+-ATPase functions as a Na^+-dependent $K^+(NH_4^+)$-ATPase in apical membranes from rat distal colon." *J. Biol. Chem.* 274:19693–19698.

Cohen, E., R. Goldshleger, A. Shainskaya, D. M. Tal, C. Ebel, M. le Maire and S. J. Karlish. 2005. "Purification of Na^+,K^+-ATPase expressed in *Pichia pastoris* reveals an essential role of phospholipid-protein interactions." *J. Biol. Chem.* 280:16610–16618.

Cougnon, M., G. Planelles, M. S. Crowson, G. E. Shull, B. C. Rossier and F. Jaisser. 1996. "The rat distal colon P-ATPase α-subunit encodes a ouabain-sensitive H^+, K^+-ATPase." *J. Biol. Chem.* 271:7277–7280.

Cougnon, M., P. Bouyer, F. Jaisser, A. Edelman and G. Planelles. 1999. "Ammonium transport by the colonic H^+-K^+-ATPase expressed in *Xenopus* oocytes." *Am. J. Physiol.* 277:C280–C287.

Cougnon, M., P. Bouyer, G. Planelles and F. Jaisser. 1998. "Does the colonic $H^+„K^+$,-ATPase also act as an Na^+,K^+-ATPase?" *Proc. Natl. Acad. Sci. U. S. A.* 95:6516–6520.

Courtois-Coutry, N., D. Roush, V. Rajendran, J. B. McCarthy, J. Geibel, M. Kashgarian and M. J. Caplan. 1997. "A tyrosine-based signal targets H^+,K^+-ATPase to a regulated compartment and is required for the cessation of gastric acid secretion." *Cell* 90:501–510.

Crambert, G., M. Fuzesi, H. Garty, S. Karlish and K. Geering. 2002. "Phospholemman (FXYD1) associates with Na^+,K^+-ATPase and regulates its transport properties." *Proc. Natl. Acad. Sci. U. S. A.* 99:11476–11481.

Crambert, G., U. Hasler, A. T. Beggah, C. Yu, N. N. Modyanov, J. D. Horisberger, L. Lelievre and K. Geering. 2000. "Transport and pharmacological properties of nine different human Na^+, K^+-ATPase isozymes." *J. Biol. Chem.* 275:1976–1986.

Crothers, J. M., Jr., S. Asano, T. Kimura, A. Yoshida, A. Wong, J. W. Kang and J. G. Forte. 2004. "Contribution of oligosaccharides to protection of the H^+,K^+-ATPase β-subunit against trypsinolysis." *Electrophoresis* 25:2586–2592.

Davenport, H. W. 1939. "Gastric carbonic anhydrase." *J. Physiol.* 97:32–43.

de Carvalho Aguiar, P., K. J. Sweadner, J. T. Penniston, J. Zaremba, L. Liu, M. Caton, G. Linazasoro, M. Borg, M. A. Tijssen, S. B. Bressman, W. B. Dobyns, A. Brashear and L. J. Ozelius. 2004. "Mutations in the Na^+,K^+-ATPase α_3-gene ATP1A3 are associated with rapid-onset dystonia parkinsonism." *Neuron* 43:169–175.

De Fusco, M., R. Marconi, L. Silvestri, L. Atorino, L. Rampoldi, L. Morgante, A. Ballabio, P. Aridon and G. Casari. 2003. "Haploinsufficiency of ATP1A2 encoding the Na^+/K^+ pump α_2-subunit associated with familial hemiplegic migraine type 2." *Nat. Genet.* 33:192–196.

Dedek, K. and S. Waldegger. 2001. "Colocalization of KCNQ1/KCNE channel subunits in the mouse gastrointestinal tract." *Pflügers Arch.* 442:896–902.

DeKoster, G. T. and A. D. Robertson. 1997. "Thermodynamics of unfolding for Kazal-type serine protease inhibitors: entropic stabilization of ovomucoid first domain by glycosylation." *Biochemistry* 36:2323–2331.

Dempski, R. E., T. Friedrich and E. Bamberg. 2005. "The β-subunit of the Na^+/K^+-ATPase follows the conformational state of the holoenzyme." *J. Gen. Physiol.* 125:505–520.

Despa, S., A.L. Tucker and D.M. Bers. 2008. "Phospholemman-mediated activation of Na^+,K^+-ATPase limits $[Na^+]_i$ and inotropic state during β-adrenergic stimulation in mouse ventricular myocytes." *Circulation* 117:1849–1855.

Donnet, C., E. Arystarkhova and K. J. Sweadner. 2001. "Thermal denaturation of the Na,K-ATPase provides evidence for α-α oligomeric interaction and γ- subunit association with the C-terminal domain." *J. Biol. Chem.* 276:7357–7365.

Dürr, K. L., K. Abe, N.N. Tavraz and T. Friedrich. 2009a. "E_2P-state stabilization by the N-terminal tail of the H^+,K^+-ATPase β-subunit is critical for efficient proton pumping under *in vivo* conditions." *J. Biol. Chem.* 284:20147–20154.

Dürr, K. L., N. N. Tavraz, D. Zimmermann, E. Bamberg and T. Friedrich. 2008. "Characterization of Na^+,K^+-ATPase and H^+,K^+-ATPase enzymes with glycosylation-deficient β-subunit variants by voltage-clamp fluorometry in *Xenopus* oocytes." *Biochemistry* 47:4288–4297.

Dürr, K. L., N. N. Tavraz, R. E. Dempski, E. Bamberg and T. Friedrich. 2009b. "Functional significance of E_2-state stabilization by specific α /β -subunit interactions of Na^+,K^+- and H^+,K^+-ATPase." *J. Biol. Chem.* 284:3842–3854.

Eakle, K. A., K. S. Kim, M. A. Kabalin and R. A. Farley. 1992. "High-affinity ouabain binding by yeast cells expressing Na^+, K^+-ATPase α-subunits and the gastric H^+, K^+-ATPase β-subunit." *Proc. Natl. Acad. Sci. U. S. A.* 89:2834–2838.

Eakle, K. A., M. A. Kabalin, S. G. Wang and R. A. Farley. 1994. "The influence of β-subunit structure on the stability of Na^+,K^+-ATPase complexes and interaction with K^+." *J. Biol. Chem.* 269:6550–6557.

Eguchi, H., S. Kaya and K. Taniguchi. 1993. "Phosphorylation of half and all sites in H^+,K^+-ATPase results in opposite changes in tryptophan fluorescence." *Biochem. Biophys. Res. Commun.* 196:294–300.

Ene, M. D., T. Khan-Daneshmend and C. J. Roberts. 1982. "A study of the inhibitory effects of SCH 28080 on gastric secretion in man." *Br. J. Pharmacol.* 76:389–391.

Fellenius, E., T. Berglindh, G. Sachs, L. Olbe, B. Elander, S. E. Sjostrand and B. Wallmark. 1981. "Substituted benzimidazoles inhibit gastric acid secretion by blocking H^+,K^+-ATPase." *Nature* 290:159–161.

Forte, J. G., G. M. Forte and P. Saltman. 1967. "K$^+$-stimulated phosphatase of microsomes from gastric mucosa." *J. Cell. Physiol.* 69:293–304.

Forte, J. G., T. M. Forte, J. A. Black, C. Okamoto and J. M. Wolosin. 1983. "Correlation of parietal cell structure and function." *J. Clin. Gastroenterol.* 5 Suppl 1:17–27.

Fröhlich, J. P., K. Taniguchi, K. Fendler, J. E. Mahaney, D. D. Thomas and R. W. Albers. 1997. "Complex kinetic behavior in the Na$^+$,K$^+$- and Ca^{2+}-ATPases. Evidence for subunit-subunit interactions and energy conservation during catalysis." *Ann. N. Y. Acad. Sci.* 834:280–296.

Gadsby, D. C., R. F. Rakowski and P. De Weer. 1993. "Extracellular access to the Na$^+$,K$^+$ pump: pathway similar to ion channel." *Science* 260:100–103.

Galmiche, J. P., S. Bruley Des Varannes, P. Ducrotté, S. Sacher-Huvelin, F. Vavasseur, A. Taccoen, P. Fiorentini and M. Homerin. 2004. "Tenatoprazole, a novel proton pump inhibitor with a prolonged plasma half-life: effects on intragastric pH and comparison with esomeprazole in healthy volunteers." *Aliment. Pharmacol. Ther.* 19:655–662.

Gassel, M. and K. Altendorf. 2001. "Analysis of KdpC of the K$^+$-transporting KdpFABC complex of *Escherichia coli*." *Eur. J. Biochem.* 268:1772–1781.

Gawenis, L.R., C Ledoussal, L.M. Judd, V. Prasad, S.L. Alper, A. Stuart-Tilley, A.L. Woo, C Grisham, L.P. Sanford, T. Doetschman, M.L. Miller and G.E. Shull. 2004. "Mice with a targeted disruption of the AE2 Cl$^-$/HCO$_3^-$ exchanger are achlorhydric." *J. Biol. Chem.* 279:30531–30539.

Gawenis, L.R, J.M Greeb, V. Prasad, C. Grisham, P. L. Sanford, T. Doetschman, A. Andringa, M.L. Miller and G.E. Shull. 2005. "Impaired gastric acid secretion in mice with a targeted disruption of the NHE4 Na$^+$/H$^+$ exchanger." *J. Biol. Chem.* 280:12781–12789.

Geering, K. 1991. "The functional role of the β-subunit in the maturation and intracellular transport of Na$^+$,K$^+$-ATPase." *FEBS Lett.* 285:189–193.

Geering, K. 2001. "The functional role of β-subunits in oligomeric P-type ATPases." *J. Bioenerg. Biomembr.* 33:425–438.

Geering, K., A. Beggah, P. Good, S. Girardet, S. Roy, D. Schaer and P. Jaunin. 1996. "Oligomerization and maturation of Na$^+$,K$^+$-ATPase: functional interaction of the cytoplasmic NH $_2$-terminus of the β-subunit with the α-subunit." *J. Cell. Biol.* 133:1193–1204.

Geering, K., I. Theulaz, F. Verrey, M. T. Hauptle and B. C. Rossier. 1989. "A role for the β-subunit in the expression of functional Na$^+$,K$^+$-ATPase in *Xenopus* oocytes." *Am. J. Physiol.* 257:C851–C858.

Geibel, S., D. Zimmermann, G. Zifarelli, A. Becker, J. B. Koenderink, Y. K. Hu, J. H. Kaplan, T. Friedrich and E. Bamberg. 2003a. "Conformational dynamics of Na$^+$/K$^+$- and H$^+$/K$^+$-ATPase probed by voltage clamp fluorometry." *Ann. N. Y. Acad. Sci.* 986:31–38.

Geibel, S., J. H. Kaplan, E. Bamberg and T. Friedrich. 2003b. "Conformational dynamics of the Na$^+$,K$^+$-ATPase probed by voltage clamp fluorometry." *Proc. Natl. Acad. Sci. U. S. A.* 100:964–969.

Golovina, V. A., H. Song, P. F. James, J. B. Lingrel and M. P. Blaustein. 2003. "Na$^+$ pump α$_2$-subunit expression modulates Ca^{2+} signaling." *Am. J. Physiol. Cell. Physiol.* 284:C475–C486.

Gottardi, C. J. and M. J. Caplan. 1993. "Molecular requirements for the cell-surface expression of multisubunit ion-transporting ATPases. Identification of protein domains that participate in Na$^+$,K$^+$-ATPase and H$^+$,K$^+$-ATPase subunit assembly." *J. Biol. Chem.* 268:14342–14347.

Grahammer, F., A. W. Herling, H. J. Lang, A. Schmitt-Graff, O. H. Wittekindt, R. Nitschke, M. Bleich, J. Barhanin and R. Warth. 2001. "The cardiac K$^+$ channel KCNQ1 is essential for gastric acid secretion." *Gastroenterology* 120:1363–1371.

Grishin, A. V. and M. J. Caplan. 1998. "ATP1AL1, a member of the non-gastric H^+,K^+-ATPase family, functions as a sodium pump." *J. Biol. Chem.* 273:27772–27778.

Harris, J. B., H. Frank and I. S. Edelman. 1958. "Effect of potassium on ion transport and bioelectric potentials of frog gastric mucosa." *Am. J. Physiol.* 195:499–504.

Hasler, U., G. Crambert, J. D. Horisberger and K. Geering. 2001. "Structural and functional features of the transmembrane domain of the Na^+,K^+-ATPase β-subunit revealed by tryptophan scanning." *J. Biol. Chem.* 276:16356–16364.

Hasler, U., P. J. Greasley, G. von Heijne and K. Geering. 2000. "Determinants of topogenesis and glycosylation of type II membrane proteins. Analysis of Na^+,K^+-ATPase $β_1$- and/or $β_3$-subunits by glycosylation mapping." *J. Biol. Chem.* 275:29011–29022.

Hasler, U., X. Wang, G. Crambert, P. Beguin, F. Jaisser, J. D. Horisberger and K. Geering. 1998. "Role of β-subunit domains in the assembly, stable expression, intracellular routing, and functional properties of Na^+,K^+-ATPase." *J. Biol. Chem.* 273:30826–30835.

Hayashi, Y., K. Kameyama, T. Kobayashi, E. Hagiwara, N. Shinji and T. Takagi. 1997. "Oligomeric structure of solubilized Na^+,K^+-ATPase linked to E_1/E_2 conformation." *Ann. N. Y. Acad. Sci.* 834:19–29.

Hayashi, Y., K. Mimura, H. Matsui and T. Takagi. 1989. "Minimum enzyme unit for Na^+/K^+-ATPase is the α/β-protomer. Determination by low-angle laser light scattering photometry coupled with high-performance gel chromatography for substantially simultaneous measurement of ATPase activity and molecular weight." *Biochim. Biophys. Acta.* 983:217–229.

Hebert, H., Y. Xian, I. Hacksell and S. Mardh. 1992. "Two-dimensional crystals of membrane-bound gastric H^+,K^+-ATPase." *FEBS Lett.* 299:159–162.

Hegyvary, C. and R. L. Post. 1971. "Binding of adenosine triphosphate to sodium and potassium ion-stimulated adenosine triphosphatase." *J. Biol. Chem.* 246:5234–5240.

Heitzmann, D., F. Grahammer, T. von Hahn, A. Schmitt-Graff, E. Romeo, R. Nitschke, U. Gerlach, H. J. Lang, F. Verrey, J. Barhanin and R. Warth. 2004. "Heteromeric KCNE2/KCNQ1 potassium channels in the luminal membrane of gastric parietal cells." *J. Physiol.* 561:547–557.

Helander, H. F., R. Leth and L. Olbe. 1986. "Stereological investigations on human gastric mucosa: I. Normal oxyntic mucosa." *Anat. Rec.* 216:373–380.

Heller, M., M. von der Ohe, R. Kleene, M. H. Mohajeri and M. Schachner. 2003. "The immunoglobulin-superfamily molecule basigin is a binding protein for oligomannosidic carbohydrates: an anti-idiotypic approach." *J. Neurochem.* 84:557–565.

Helmich-de Jong, M. L., S. E. van Emst-de Vries, H. G. Swarts, F. M. Schuurmans Stekhoven and J. J. de Pont. 1986. "Presence of a low-affinity nucleotide binding site on the H^+,K^+-ATPase phosphoenzyme." *Biochim. Biophys. Acta* 860:641–649.

Helmich-de Jong, M. L., S. E. van Emst-de Vries, J. J. De Pont, F. M. Schuurmans Stekhoven and S. L. Bonting. 1985. "Direct evidence for an ADP-sensitive phosphointermediate of H^+,K^+-ATPase." *Biochim. Biophys. Acta* 821:377–383.

Hersey, S. J., L. Steiner, J. Mendlein, E. Rabon and G. Sachs. 1988. "SCH 28080 prevents omeprazole inhibition of the gastric H^+,K^+-ATPase." *Biochim. Biophys. Acta* 956:49–57.

Holmgren, M., J. Wagg, F. Bezanilla, R. F. Rakowski, P. De Weer and D. C. Gadsby. 2000. "Three distinct and sequential steps in the release of sodium ions by the Na^+,K^+-ATPase." *Nature* 403:898–901.

Horisberger, J. D., P. Jaunin, M. A. Reuben, L. S. Lasater, D. C. Chow, J. G. Forte, G. Sachs, B. C. Rossier and K. Geering. 1991. "The H^+,K^+-ATPase β-subunit can act as a surrogate for the β-subunit of Na^+,K^+-pumps." *J. Biol. Chem* 266:19131–19134.

Hu, Y. K. and J. H. Kaplan. 2000. "Site-directed chemical labeling of extracellular loops in a membrane protein. The topology of the Na^+,K^+-ATPase α-subunit." *J. Biol. Chem* 275:19185–19191.

Im, W. B., D. P. Blakeman, J. Mendlein and G. Sachs. 1984. "Inhibition of H^+, K^+-ATPase and H^+ accumulation in hog gastric membranes by trifluoperazine, verapamil and 8-(N,N-diethylamino)octyl-3,4,5-trimethoxybenzoate." *Biochim. Biophys. Acta* 770:65–72.

Imperiali, B. and S. E. O'Connor. 1999. "Effect of N-linked glycosylation on glycopeptide and glycoprotein structure." *Curr. Opin. Chem. Biol.* 3:643–649.

Ivanov, A.V., N.N. Modyanov and A. Askari. 2002. "Role of the self-association of β-subunits in the oligomeric structure of Na^+,K^+-ATPase." *Biochem. J.* 364:293–299.

Jaisser, F., J. D. Horisberger, K. Geering and B. C. Rossier. 1993. "Mechanisms of urinary K^+ and H^+ excretion: primary structure and functional expression of a novel H^+,K^+-ATPase." *J. Cell. Biol.* 123:1421–1429.

Jaisser, F., P. Jaunin, K. Geering, B. C. Rossier and J. D. Horisberger. 1994. "Modulation of the Na^+,K^+-pump function by β-subunit isoforms." *J. Gen. Physiol.* 103:605–623.

James, P., M. Inui, M. Tada, M. Chiesi and E. Carafoli. 1989. "Nature and site of phospholamban regulation of the Ca^{2+} pump of sarcoplasmic reticulum." *Nature* 342:90–92.

Jaunin, P., F. Jaisser, A. T. Beggah, K. Takeyasu, P. Mangeat, B. C. Rossier, J. D. Horisberger and K. Geering. 1993. "Role of the transmembrane and extracytoplasmic domain of β-subunits in subunit assembly, intracellular transport, and functional expression of Na^+,K^+-pumps." *J. Cell. Biol.* 123:1751–1759.

Juhaszova, M. and M. P. Blaustein. 1997. "Na^+ pump low and high ouabain affinity α-subunit isoforms are differently distributed in cells." *Proc. Natl. Acad. Sci. U. S. A.* 94:1800–1805.

Kaminski, J. J., J. A. Bristol, C. Puchalski, R. G. Lovey, A. J. Elliott, H. Guzik, D. M. Solomon, D. J. Conn, M. S. Domalski and S. C. Wong. 1985. "Antiulcer agents. 1. Gastric antisecretory and cytoprotective properties of substituted imidazo[1,2-a]pyridines." *J. Med. Chem.* 28:876–892.

Kamsteeg, E. J. and P. M. Deen. 2000. "Importance of aquaporin-2 expression levels in genotype-phenotype studies in nephrogenic diabetes insipidus." *Am. J. Physiol. Renal. Physiol.* 279:F778–F784.

Kamsteeg, E. J. and P. M. Deen. 2001. "Detection of aquaporin-2 in the plasma membranes of oocytes: a novel isolation method with improved yield and purity." *Biochem. Biophys. Res. Commun.* 282:683–690.

Kaplan, J. H. 1982. "Sodium pump-mediated ATP:ADP exchange. The sided effects of sodium and potassium ions." *J. Gen. Physiol.* 80:915–937.

Kaplan, J. H. and R. J. Hollis. 1980. "External Na^+ dependence of ouabain-sensitive ATP:ADP exchange initiated by photolysis of intracellular caged-ATP in human red cell ghosts." *Nature* 288:587–589.

Katoh, Y. and M. Katoh. 2004. "Identification and characterization of CDC50A, CDC50B and CDC50C genes in silico." *Oncol. Rep.* 12:939–943.

Kawamura, M., K. Ohmizo, M. Morohashi and K. Nagano. 1985. "Protective effect of Na^+ and K^+ against inactivation of Na^+,K^+-ATPase by high concentrations of 2-mercaptoethanol at high temperatures." *Biochim. Biophys. Acta* 821:115–120.

Kaya, S., K. Abe, K Taniguchi, M. Yazawa, T. Katoh, M. Kikumoto, K. Oiwa and Y. Hayashi. 2003. "Oligomeric structure of P-type ATPases observed by single molecule detection technique." *Ann. N. Y. Acad. Sci.* 986:278–280.

Keeling, D. J., S. M. Laing and J. Senn-Bilfinger. 1988. "SCH 28080 is a lumenally acting, K^+-site inhibitor of the gastric H^+,K^+-ATPase." *Biochem. Pharmacol.* 37:2231–2236.

Kühlbrandt, W. 2004. "Biology, structure and mechanism of P-type ATPases." *Nat. Rev. Mol. Cell. Biol.* 5:282–295.

Kitamura, N., M. Ikekita, T. Sato, Y. Akimoto, Y. Hatanaka, H. Kawakami, M. Inomata and K. Furukawa. 2005. "Mouse Na^+,K^+-ATPase β_1-subunit has a K^+-dependent cell adhesion activity for β-GlcNAc-terminating glycans." *Proc. Natl. Acad. Sci. U. S. A.* 102:2796–27801.

Klaassen, C. H., J. A. Fransen, H. G. Swarts and J. J. De Pont. 1997. "Glycosylation is essential for biosynthesis of functional gastric H^+,K^+-ATPase in insect cells." *Biochem. J.* 321 (Pt 2):419–424.

Klodos, I. and III. Forbush. 1988. "Rapid conformational changes of the Na,K pump revealed by a fluoresecent dye, RH-160." *J. Gen. Physiol.* 92:46a.

Kobayashi, T., Y. Tahara, H. Takenaka, K. Mimura and Y. Hayashi. 2007. "Na^+- and K^+-dependent oligomeric interconversion among α/β-protomers, diprotomers and higher oligomers in solubilized Na^+,K^+-ATPase." *J. Biochem.* 142:157–173.

Koenderink, J. B., H. P. Hermsen, H. G. Swarts, P. H. Willems and J. J. De Pont. 2000. "High-affinity ouabain binding by a chimeric gastric H^+,K^+-ATPase containing transmembrane hairpins M3-M4 and M5-M6 of the α_1-subunit of rat Na^+,K^+-ATPase." *Proc. Natl. Acad. Sci. U. S. A.* 97:11209–11214.

Koenderink, J. B., S. Geibel, E. Grabsch, J. J. De Pont, E. Bamberg and T. Friedrich. 2003. "Electrophysiological analysis of the mutated Na^+,K^+-ATPase cation binding pocket." *J. Biol. Chem.* 278:51213–51222.

Koster, J. C., G. Blanco and R. W. Mercer. 1995. "A cytoplasmic region of the Na^+,K^+-ATPase α-subunit is necessary for specificα/α association." *J. Biol. Chem.* 270:14332–14339.

Kozak, M. 1987. "An analysis of 5'-noncoding sequences from 699 messenger RNAs." *Nucleic. Acids. Res.* 15:8125–8148.

Lachman, L. and C. W. Howden. 2000. "Twenty-four-hour intragastric pH: tolerance within 5 days of continuous ranitidine administration." *Am. J. Gastroenterol.* 95:57–61.

Laemmli, U. K. 1970. "Cleavage of structural proteins during the assembly of the head of bacteriophage T4." *Nature* 227:680–685.

Lambrecht, N. W., I. Yakubov, D. Scott and G. Sachs. 2005. "Identification of the K^+ efflux channel coupled to the gastric H^+,K^+-ATPase during acid secretion." *Physiol. Genomics* 21:81–91.

Lambrecht, N.W.G, I. Yakubov, C. Zer and G. Sachs. 2006. "Transcriptomes of purified gastric ECL and parietal cells: identification of a novel pathway regulating acid secretion." *Physiol. Genomics* 25:153–165.

Lang, F., G. L. Busch and H. Völkl. 1998. "The diversity of volume regulatory mechanisms." *Cell. Physiol. Biochem.* 8:1–45.

Laughery, M., M. Todd and J. H. Kaplan. 2004. "Oligomerization of the Na,K-ATPase in cell membranes." *J. Biol. Chem.* 279:36339–36348.

Lee, M. P., J. D. Ravenel, R. J. Hu, L. R. Lustig, G. Tomaselli, R. D. Berger, S. A. Brandenburg, T. J. Litzi, T. E. Bunton, C. Limb, H. Francis, M. Gorelikow, H. Gu, K. Washington, P. Argani, J. R. Goldenring, R. J. Coffey and A. P. Feinberg. 2000. "Targeted disruption of the Kvlqt1 gene causes deafness and gastric hyperplasia in mice." *J. Clin. Invest.* 106:1447–1455.

Lenoir, G., P. Williamson, C. F. Puts and J. C. M. Holthuis. 2009. "Cdc50p Plays a Vital Role in the ATPase Reaction Cycle of the Putative Aminophospholipid Transporter DRS2P." *J. Biol. Chem.* (published online).

Lenoir, G., P. Williamson and J.C.M. Holthuis. 2007. "On the origin of lipid asymmetry: the flip side of ion transport." *Curr. Opin. Chem. Biol.* 11:654–661.

Li, C., O. Capendeguy, K. Geering and J. D. Horisberger. 2005. "A third Na^+-binding site in the sodium pump." *Proc. Natl. Acad. Sci. U. S. A.* pp. 12706–12711.

Lian, W. N., T. W. Wu, R. L. Dao, Y. J. Chen and C. H. Lin. 2006. "Deglycosylation of Na^+/K^+-ATPase causes the basolateral protein to undergo apical targeting in polarized hepatic cells." *J. Cell. Sci.* 119:11–22.

Lindberg, P., P. Nordberg, T. Alminger, A. Brandstrom and B. Wallmark. 1986. "The mechanism of action of the gastric acid secretion inhibitor omeprazole." *J. Med. Chem.* 29:1327–1329.

Long, J. F., P. J. Chiu, M. J. Derelanko and M. Steinberg. 1983. "Gastric antisecretory and cytoprotective activities of SCH 28080." *J. Pharmacol. Exp. Ther.* 226:114–120.

Loomes, K. M., H. E. Senior, P. M. West and A. M. Roberton. 1999. "Functional protective role for mucin glycosylated repetitive domains." *Eur. J. Biochem.* 266:105–111.

Lorentzon, P., B. Eklundh, A. Brandstrom and B. Wallmark. 1985. "The mechanism for inhibition of gastric H^+,K^+-ATPase by omeprazole." *Biochim. Biophys. Acta* 817:25–32.

Lorentzon, P., G. Sachs and B. Wallmark. 1988. "Inhibitory effects of cations on the gastric H^+,K^+-ATPase. A potential-sensitive step in the K^+ limb of the pump cycle." *J. Biol. Chem.* 263:10705–10710.

Lorenz, C., M. Pusch and T. J. Jentsch. 1996. "Heteromultimeric CLC chloride channels with novel properties." *Proc. Natl. Acad. Sci. U. S. A.* 93:13362–13366.

Läuger, P. 1991. *Electrogenic Ion Pumps*. Sinauer, Sunderland, MA.

Lutsenko, S. and J. H. Kaplan. 1993. "An essential role for the extracellular domain of the Na^+,K^+-ATPase β-subunit in cation occlusion." *Biochemistry* 32:6737–6743.

Ma, T., A. Frigeri, H. Hasegawa and A. S. Verkman. 1994. "Cloning of a water channel homolog expressed in brain meningeal cells and kidney collecting duct that functions as a stilbene-sensitive glycerol transporter." *J Biol Chem* 269:21845–21849.

MacLennan, D.H. and E.G. Kranias. 2003. "Phospholamban: a crucial regulator of cardiac contractility." *Nat. Rev. Mol. Cell. Biol.* 4:566–577.

Marquardt, T. and A. Helenius. 1992. "Misfolding and aggregation of newly synthesized proteins in the endoplasmic reticulum." *J. Cell. Biol.* 117:505–513.

Marshall, B. J. and J. R. Warren. 1984. "Unidentified curved bacilli in the stomach of patients with gastritis and peptic ulceration." *Lancet* 1:1311–1315.

Mathews, P. M., D. Claeys, F. Jaisser, K. Geering, J. D. Horisberger, J. P. Kraehenbuhl and B. C. Rossier. 1995. "Primary structure and functional expression of the mouse and frog β-subunit of the gastric H^+,K^+-ATPase." *Am. J. Physiol.* 268:C1207–1214.

Maunsbach, A. B., E. Skriver and H. Hebert. 1991. "Two-dimensional crystals and three-dimensional structure of Na^+,K^+-ATPase analyzed by electron microscopy." *Soc. Gen. Physiol. Ser.* 46:159–172.

McDonough, A. A., K. Geering and R. A. Farley. 1990. "The sodium pump needs its β-subunit." *FASEB J* . 4:1598–1605.

Meier, S., N. N. Tavraz, K. L. Dürr and T. Friedrich. 2009. "Hyperpolarisation-activated Inward Na^+ Leakage Currents Caused by Deletion or Mutation of Carboxy-terminal Tyrosines of the Na^+,K^+-ATPase α_2-Subunit." *J. Gen. Physiol.* (submitted).

Meij, I. C., J. B. Koenderink, H. van Bokhoven, K. F. Assink, W. T. Groenestege, J. J. de Pont, R. J. Bindels, L. A. Monnens, L. P. van den Heuvel and N. V. Knoers. 2000. "Dominant isolated renal magnesium loss is caused by misrouting of the Na^+,K^+-ATPase γ-subunit." *Nat. Genet.* 26:265–266.

Meij, I. C., K. Saar, L. P. van den Heuvel, G. Nuernberg, M. Vollmer, F. Hildebrandt, A. Reis, L. A. Monnens and N. V. Knoers. 1999. "Hereditary isolated renal magnesium loss maps to chromosome 11q23." *Am. J. Hum. Genet.* 64:180–188.

Melle-Milovanovic, D., M. Milovanovic, S. Nagpal, G. Sachs and J. M. Shin. 1998. "Regions of association between the α- and the β-subunit of the gastric H^+,K^+-ATPase." *J. Biol. Chem.* 273:11075–11081.

Mimura, K., Y. Tahara, N. Shinji, E. Tokuda, H. Takenaka and Y. Hayashi. 2008. "Isolation of stable $(\alpha/\beta)_4$-Tetraprotomer from Na^+,K^+-ATPase solubilized in the presence of short-chain fatty acids." *Biochemistry* 47:6039–6051.

Misaka, T., K. Abe, K. Iwabuchi, Y. Kusakabe, M. Ichinose, K. Miki, Y. Emori and S. Arai. 1996. "A water channel closely related to rat brain aquaporin 4 is expressed in acid- and pepsinogen-secretory cells of human stomach." *FEBS Lett.* 381:208–212.

Mizukawa, Y., T. Nishizawa, T. Nagao, K. Kitamura and T Urushidani. 2002. "Cellular distribution of parchorin, a chloride intracellular channel-related protein, in various tissues." *Am. J. Physiol. Cell. Physiol.* 282:C786–C795.

Mori, Y., K. Fukuma, Y. Adachi, K. Shigeta, R. Kannagi, H. Tanaka, M. Sakai, K. Kuribayashi, H. Uchino and T. Masuda. 1989. "Parietal cell autoantigens involved in neonatal thymectomy-induced murine autoimmune gastritis. Studies using monoclonal autoantibodies." *Gastroenterology* 97:364–375.

Morii, M., Y. Hayata, K. Mizoguchi and N. Takeguchi. 1996. "Oligomeric regulation of gastric H^+,K^+-ATPase." *J. Biol. Chem.* 271:4068–4072.

Morth, J. P., B. P. Pedersen, M. S. Toustrup-Jensen, T. L. Sorensen, J. Petersen, J. P. Andersen, B. Vilsen and P. Nissen. 2007. "Crystal structure of the sodium-potassium pump." *Nature* 450:1043–1049.

Moskowitz, M.A., H. Bolay and T Dalkara. 2004. "Deciphering migraine mechanisms: clues from familial hemiplegic migraine genotypes." *Ann. Neurol.* 55:276–280.

Muallem, S., C. Burnham, D. Blissard, T. Berglindh and G. Sachs. 1985. "Electrolyte transport across the basolateral membrane of the parietal cells." *J. Biol. Chem.* 260:6641–6653.

Nguyen, N. V., P. A. Gleeson, N. Courtois-Coutry, M. J. Caplan and I. R. Van Driel. 2004. "Gastric parietal cell acid secretion in mice can be regulated independently of H^+,K^+- ATPase endocytosis." *Gastroenterology* 127:145–154.

Noguchi, S., M. Mishina, M. Kawamura and S. Numa. 1987. "Expression of functional Na^+, K^+-ATPase from cloned cDNAs." *FEBS Lett.* 225:27–32.

Ogata, T. and Y. Yamasaki. 2000. "Scanning EM of resting gastric parietal cells reveals a network of cytoplasmic tubules and cisternae connected to the intracellular canaliculus." *Anat. Rec.* 258:15–24.

Ogawa, H., T. Shinoda, F. Cornelius and C. Toyoshima. 2009. "Crystal structure of the sodium-potassium pump (Na$^+$,K$^+$-ATPase) with bound potassium and ouabain." *Proc. Natl. Acad. Sci. U. S. A.*.

Olesen, C., M. Picard, A. M. Winther, C. Gyrup, J. P. Morth, C. Oxvig, J. V. Moller and P. Nissen. 2007. "The structural basis of calcium transport by the calcium pump." *Nature* 450:1036–1042.

Olesen, C., T. L. Sorensen, R. C. Nielsen, J. V. Moller and P. Nissen. 2004. "Dephosphorylation of the calcium pump coupled to counterion occlusion." *Science* 306:2251–2255.

Orlowski, J., R. A. Kandasamy and G. E. Shull. 1992. "Molecular cloning of putative members of the Na/H exchanger gene family. cDNA cloning, deduced amino acid sequence, and mRNA tissue expression of the rat Na/H exchanger NHE-1 and two structurally related proteins." *J. Biol. Chem.* 267:9331–9339.

Pachence, J. M., I. S. Edelman and B. P. Schoenborn. 1987. "Low-angle neutron scattering analysis of Na$^+$,K$^+$-ATPase in detergent solution." *J. Biol. Chem.* 262:702–709.

Palasis, M., T. A. Kuntzweiler, J. M. Arguello and J. B. Lingrel. 1996. "Ouabain interactions with the H5-H6 hairpin of the Na$^+$,K$^+$-ATPase reveal a possible inhibition mechanism via the cation binding domain." *J. Biol. Chem.* 271:14176–14182.

Paradiso, A. M., R. Y. Tsien and T. E. Machen. 1987. "Digital image processing of intracellular pH in gastric oxyntic and chief cells." *Nature* 325:447–450.

Petrovic, S., X. Ju, S. Barone, U. Seidler, S. L. Alper, H. Lohi, J. Kere and M. Soleimani. 2003. "Identification of a basolateral Cl$^-$/HCO$_3^-$ exchanger specific to gastric parietal cells." *Am. J. Physiol. Gastrointest. Liver Physiol.* 284:G1093–G1103.

Pilotelle-Bunner, A., J. Matthews, Cornelius F., H. J. Apell, P. Sebban and R. Clarke. 2008. "ATP Binding Equilibria of the Na$^+$,K$^+$-ATPase." *Biochemistry* 47:13103–13114.

Polvani, C., G. Sachs and R. Blostein. 1989. "Sodium ions as substitutes for protons in the gastric H$^+$,K$^+$-ATPase." *J. Biol. Chem.* 264:17854–17859.

Post, R. L., C. Hegyvary and S. Kume. 1972. "Activation by adenosine triphosphate in the phosphorylation kinetics of sodium and potassium ion transport adenosine triphosphatase." *J. Biol. Chem.* 247:6530–6540.

Poulsen, L. R., López-Marqués R. L., McDowell S.C., J. Okkeri, D. Licht, A. Schulz, T. Pomorski, J.F Harper and M.G. Palmgren. 2008a. "The Arabidopsis P4-ATPase ALA3 localizes to the golgi and requires a β-subunit to function in lipid translocation and secretory vesicle formation." *Plant Cell* 20:658–676.

Poulsen, L. R., R. L. López-Marqués and M. G. Palmgren. 2008b. "Flippases: still more questions than answers." *Cell Mol Life Sci* 65:3119–3125.

Price, E. M. and J. B. Lingrel. 1988. "Structure-function relationships in the Na$^+$,K$^+$-ATPase α-subunit: site-directed mutagenesis of glutamine-111 to arginine and asparagine-122 to aspartic acid generates a ouabain-resistant enzyme." *Biochemistry* 27:8400–8408.

Puts, C. F. and J. C. M. Holthuis. 2009. "Mechanism and significance of P(4) ATPase-catalyzed lipid transport: Lessons from a Na$^+$,K$^+$-pump." *Biochim. Biophys. Acta.*.

Qiu, L. Y., E. Krieger, G. Schaftenaar, H. G. Swarts, P. H. Willems, J. J. De Pont and J. B. Koenderink. 2005. "Reconstruction of the complete ouabain-binding pocket of Na$^+$,K$^+$-ATPase in gastric H$^+$,K$^+$-ATPase by substitution of only seven amino acids." *J. Biol. Chem.* 280:32349–32355.

Qiu, L. Y., H. G. Swarts, E. C. Tonk, P. H. Willems, J. B. Koenderink and J. J. De Pont. 2006. "Conversion of the low affinity ouabain-binding site of non-gastric H^+,K^+-ATPase into a high affinity binding site by substitution of only five amino acids." *J. Biol. Chem.* 281:13533–14539.

Rabon, E. C., S. Bassilian, G. Sachs and S. J. Karlish. 1990. "Conformational transitions of the H^+,K^+-ATPase studied with sodium ions as surrogates for protons." *J. Biol. Chem.* 265:19594–19599.

Rajendran, V. M., P. Sangan, J. Geibel and H. J. Binder. 2000. "Ouabain-sensitive H^+,K^+-ATPase functions as Na^+,K^+-ATPase in apical membranes of rat distal colon." *J. Biol. Chem.* 275:13035–13040.

Rakowski, R. F., D. C. Gadsby and P. De Weer. 1997. "Voltage dependence of the Na^+,K^+ pump." *J. Membr. Biol.* 155:105–112.

Rakowski, R. F., L. A. Vasilets, J. LaTona and W. Schwarz. 1991. "A negative slope in the current-voltage relationship of the Na^+,K^+ pump in *Xenopus* oocytes produced by reduction of external $[K^+]$." *J. Membr. Biol.* 121:177–187.

Ray, T. K. and J. Nandi. 1985. "Modulation of gastric H^+,K^+-transporting ATPase function by sodium." *FEBS Lett.* 185:24–28.

Reenstra, W. W. and J. G. Forte. 1990. "Characterization of K^+ and Cl^- conductances in apical membrane vesicles from stimulated rabbit oxyntic cells." *Am. J. Physiol.* 259:G850–G858.

Rehm, W. S. 1965. "Electrophysiology of the gastric mucosa in chloride-free solutions." *Fed. Proc.* 24:1387–1395.

Renaud, K. J., E. M. Inman and D. M. Fambrough. 1991. "Cytoplasmic and transmembrane domain deletions of Na^+,K^+-ATPase β-subunit. Effects on subunit assembly and intracellular transport." *J. Biol. Chem.* 266:20491–20497.

Repke, K. R. and R. Schön. 1973. "Flip-flop model of Na^+,K^+-ATPase function." *Acta. Biol. Med. Ger.* 31:Suppl:K19–Suppl:K30.

Riant, F., M. De Fusco, P. Aridon, A. Ducros, C. Ploton, F. Marchelli, J. Maciazek, M. G. Bousser, G. Casari and E. Tournier-Lasserve. 2005. "ATP1A2 mutations in 11 families with familial hemiplegic migraine." *Hum. Mutat.* 26:281.

Robinson, J. D. 1967. "Kinetic studies on a brain microsomal adenosine triphosphatase. Evidence suggesting conformational changes." *Biochemistry* 6:3250–3258.

Roepke, T. K., A. Anantharam, P. Kirchhoff, S. M. Busque, J. B. Young, J. P. Geibel, D. J. Lerner and G. W. Abbott. 2006. "The KCNE2 potassium channel ancillary subunit is essential for gastric acid secretion." *J. Biol. Chem.* 281:23740–23747.

Rossmann, H., T. Sonnentag, A. Heinzmann, B. Seidler, O. Bachmann, D. Vieillard-Baron, M. Gregor and U. Seidler. 2001. "Differential expression and regulation of Na^+/H^+ exchanger isoforms in rabbit parietal and mucous cells." *Am. J. Physiol. Gastrointest. Liver Physiol.* 281:G447–G458.

Roush, D. L., C. J. Gottardi, H. Y. Naim, M. G. Roth and M. J. Caplan. 1998. "Tyrosine-based membrane protein sorting signals are differentially interpreted by polarized Madin-Darby canine kidney and LLC-PK1 epithelial cells." *J. Biol. Chem.* 273:26862–26869.

Rudd, P. M., H. C. Joao, E. Coghill, P. Fiten, M. R. Saunders, G. Opdenakker and R. A. Dwek. 1994. "Glycoforms modify the dynamic stability and functional activity of an enzyme." *Biochemistry* 33:17–22.

Sachs, G. 2003. "Physiology of the parietal cell and therapeutic implications." *Pharmacotherapy* 23:68S–73S.

Sachs, G., H. H. Chang, E. Rabon, R. Schackman, M. Lewin and G. Saccomani. 1976. "A nonelectrogenic H^+ pump in plasma membranes of hog stomach." *J. Biol. Chem.* 251:7690–7698.

Sachs, G., J. M. Shin and C. W. Howden. 2006. "Review article: the clinical pharmacology of proton pump inhibitors." *Aliment. Pharmacol. Ther.* 23 Suppl 2:2–8.

Sachs, G., J. M. Shin, O. Vagin, N. Lambrecht, I. Yakubov and K. Munson. 2007. "The gastric H^+,K^+-ATPase as a drug target: past, present, and future." *J. Clin. Gastroentero.l* 41:S226–S242.

Sagar, A. and R. F. Rakowski. 1994. "Access channel model for the voltage dependence of the forward-running Na^+, K^+ pump." *J. Gen. Physiol.* 103:869–893.

Saito, K, K. Fujimura-Kamada, N. Furuta, U. Kato, M. Umeda and K. Tanaka. 2004. "Cdc50p, a protein required for polarized growth, associates with the Drs2p P-type ATPase implicated in phospholipid translocation in *Saccharomyces cerevisiae*." *Mol. Biol .Cell* 15:3418–3432.

Schmitt, J.P., M. Kamisago, M. Asahi, G.H. Li, F. Ahmad, U. Mende, E.G. Kranias, Seidman J.G. MacLennan, D.H and C.E. Seidman. 2003. "Dilated cardiomyopathy and heart failure caused by a mutation in phospholamban." *Science* 299:1410–1413.

Segall, L., S. E. Daly and R. Blostein. 2001. "Mechanistic basis for kinetic differences between the rat α_1-, α_2-, and α_3-isoforms of the Na^+,K^+-ATPase." *J. Biol. Chem.* 276:31535–31541.

Shainskaya, A. and S. J. Karlish. 1996. "Chymotryptic digestion of the cytoplasmic domain of the β-subunit of Na^+,K^+-ATPase alters kinetics of occlusion of Rb^+ ions." *J. Biol. Chem.* 271:10309–10316.

Shin, J. M., G. Grundler, J. Senn-Bilfinger, W. A. Simon and G. Sachs. 2005. "Functional consequences of the oligomeric form of the membrane-bound gastric H^+,K^+-ATPase." *Biochemistry* 44:16321–16332.

Shin, J. M. and G. Sachs. 1996. "Dimerization of the gastric H^+, K^+-ATPase." *J. Biol. Chem.* 271:1904–1908.

Shin, J. M. and G. Sachs. 2004. "Differences in binding properties of two proton pump inhibitors on the gastric H^+,K^+-ATPase *in vivo*." *Biochem. Pharmacol.* 68:2117–2127.

Shin, J. M., M. Besancon, A. Simon and G. Sachs. 1993. "The site of action of pantoprazole in the gastric H^+,K^+-ATPase." *Biochim. Biophys. Acta* 1148:223–233.

Shin, J. M., M. Homerin, F. Domagala, H. Ficheux and G. Sachs. 2006. "Characterization of the inhibitory activity of tenatoprazole on the gastric H^+,K^+ -ATPase *in vitro* and *in vivo*." *Biochem. Pharmacol.* 71:837–849.

Shinoda, T., H. Ogawa, F. Cornelius and C. Toyoshima. 2009. "Crystal structure of the sodium-potassium pump at 2.4 Å resolution." *Nature* 459:446–450.

Simmerman, H. K., J. H. Collins, J. L. Theibert, A. D. Wegener and L. R. Jones. 1986. "Sequence analysis of phospholamban. Identification of phosphorylation sites and two major structural domains." *J. Biol. Chem.* 261:13333–13341.

Skou, J. C. 1957. "The influence of some cations on an adenosine triphosphatase from peripheral nerves." *Biochim Biophys Acta* 23:394–401.

Sonnentag, T., W. K. Siegel, O. Bachmann, H. Rossmann, A. Mack, H. J. Wagner, M. Gregor and U. Seidler. 2000. "Agonist-induced cytoplasmic volume changes in cultured rabbit parietal cells." *Am. J. Physiol. Gastrointest. Liver. Physiol.* 279:G40–G48.

Stein, W. D., W. R. Lieb, S. J. Karlish and Y. Eilam. 1973. "A modle for active transport of sodium and potassium ions as mediated by a tetrameric enzyme." *Proc Natl Acad Sci U S A* 70:275–278.

Stengelin, M., K. Fendler and E. Bamberg. 1993. "Kinetics of transient pump currents generated by the H^+,K^+-ATPase after an ATP concentration jump." *J. Membr. Biol.* 132:211–227.

Stuart-Tilley, A., C. Sardet, J. Pouyssegur, M. A. Schwartz, D. Brown and S. L. Alper. 1994. "Immunolocalization of anion exchanger AE2 and cation exchanger NHE-1 in distinct adjacent cells of gastric mucosa." *Am. J. Physiol.* 266:C559–C568.

Sugai, N. and S. Ito. 1980. "Carbonic anhydrase, ultrastructural localization in the mouse gastric mucosa and improvements in the technique." *J. Histochem. Cytochem.* 28:511–525.

Sun, Y. and Jr. Ball, W. J. 1994. "Identification of antigenic sites on the Na^+,K^+-ATPase β-subunit: their sequences and the effects of thiol reduction upon their structure." *Biochim. Biophys. Acta* 1207:236–248.

Swarts, H. G., C. H. Klaassen, F. M. Schuurmans Stekhoven and J. J. De Pont. 1995. "Sodium acts as a potassium analog on gastric H^+,K^+-ATPase." *J. Biol. Chem.* 270:7890–7895.

Sweadner, K. J. and E. Rael. 2000. "The FXYD gene family of small ion transport regulators or channels: cDNA sequence, protein signature sequence, and expression." *Genomics* 68:41–56.

Takahashi, Y., H. Sakai, M. Kuragari, T. Suzuki, K. Tauchi, T. Minamimura, K. Tsukada, S. Asano and N. Takeguchi. 2002. "Expression of ATP1AL1, a non-gastric proton pump, in human colorectum." *Jpn J. Physiol.* 52:317–321.

Takeda, K., S. Noguchi, A. Sugino and M. Kawamura. 1988. "Functional activity of oligosaccharide-deficient Na^+,K^+-ATPase expressed in *Xenopus* oocytes." *FEBS Lett.* 238:201–204.

Tamkun, M. M. and D. M. Fambrough. 1986. "The Na^+,K^+-ATPase of chick sensory neurons. Studies on biosynthesis and intracellular transport." *J. Biol. Chem.* 261:1009–1019.

Taniguchi, K., S. Kaya, K. Abe and S. Mardh. 2001. "The oligomeric nature of Na^+,K^+-transport ATPase." *J. Biochem.* 129:335–342.

Taniguchi, K., S. Kaya, T. Yokoyama and K. Abe. 1999. "Tetraprotomeric hypothesis of Na^+,K^+-ATPase." *Nippon Yakurigaku Zasshi* 114:179–184.

Tanzi, R. E., K. Petrukhin, I. Chernov, J. L. Pellequer, W. Wasco, B. Ross, D. M. Romano, E. Parano, L. Pavone and L. M. Brzustowicz. 1993. "The Wilson disease gene is a copper transporting ATPase with homology to the Menkes disease gene." *Nat. Genet.* 5:344–350.

Tavraz, N. N., K. L. Dürr, J. B. Koenderink, T. Freilinger, E. Bamberg, M. Dichgans and T. Friedrich. 2009. "Impaired plasma membrane targeting or protein stability by certain ATP1A2 mutations identified in sporadic or familial hemiplegic migraine." *Channels (Austin)* 3:82–87.

Tavraz, N. N., T. Friedrich, K. L. Dürr, J. B. Koenderink, E. Bamberg, T. Freilinger and M. Dichgans. 2008. "Diverse functional consequences of mutations in the Na^+/K^+-ATPase α_2-subunit causing familial hemiplegic migraine type 2." *J. Biol. Chem.* 283:31097–31106.

Tüchsen, E. and C. Woodward. 1987. "Assignment of asparagine-44 side-chain primary amide ^{1}H NMR resonances and the peptide amide N^1H resonance of glycine-37 in basic pancreatic trypsin inhibitor." *Biochemistry* 26:1918–1925.

Thangarajah, H., A. Wong, D. C. Chow, Jr. Crothers, J. M. and J. G. Forte. 2002. "Gastric H^+,K^+-ATPase and acid-resistant surface proteins." *Am. J. Physiol. Gastrointest. Liver Physiol.* 282:G953–G961.

Therien, A. G., H. X. Pu, S. J. Karlish and R. Blostein. 2001. "Molecular and functional studies of the γ-subunit of the sodium pump." *J. Bioenerg. Biomembr.* 33:407–414.

Therien, A. G., R. Goldshleger, S. J. Karlish and R. Blostein. 1997. "Tissue-specific distribution and modulatory role of the γ-subunit of the Na$^+$,K$^+$-ATPase." *J. Biol. Chem.* 272:32628–32634.

Therien, A. G., S. J. Karlish and R. Blostein. 1999. "Expression and functional role of the γ-subunit of the Na$^+$, K$^+$-ATPase in mammalian cells." *J. Biol. Chem.* 274:12252–12256.

Ting-Beall, H. P., H. C. Beall, D. F. Hastings, M. L. Friedman and W. J. Ball. 1990. "Identification of monoclonal antibody binding domains of Na$^+$,K$^+$-ATPase by immunoelectron microscopy." *FEBS Lett.* 265:121–125.

Toustrup-Jensen, M. S., R. Holm, A. P. Einholm, V. Rodacker Schack, J. P. Morth, P. Nissen, J. P. Andersen and B. Vilsen. 2009. "The C-terminus of Na$^+$,K$^+$-ATPase controls Na$^+$ affinity on both sides of the membrane through Arg935." *J. Biol. Chem.* (published online).

Toyoshima, C. and H. Nomura. 2002. "Structural changes in the calcium pump accompanying the dissociation of calcium." *Nature* 418:605–611.

Toyoshima, C., H. Nomura and T. Tsuda. 2004. "Lumenal gating mechanism revealed in calcium pump crystal structures with phosphate analogues." *Nature* 432:361–368.

Toyoshima, C., M. Nakasako, H. Nomura and H. Ogawa. 2000. "Crystal structure of the calcium pump of sarcoplasmic reticulum at 2.6 Å resolution." *Nature* 405:647–655.

Toyoshima, C. and T. Mizutani. 2004. "Crystal structure of the calcium pump with a bound ATP analogue." *Nature* 430:529–535.

Tsuda, T., S. Kaya, T. Yokoyama, Y. Hayashi and K. Taniguchi. 1998. "ATP and acetyl phosphate induces molecular events near the ATP binding site and the membrane domain of Na$^+$,K$^+$-ATPase. The tetrameric nature of the enzyme." *J. Biol. Chem.* 273:24339–24345.

Tyagarajan, K., R. R. Townsend and J. G. Forte. 1996. "The β-subunit of the rabbit H$^+$,K$^+$-ATPase:a glycoprotein with all terminal lactosamine units capped with α-linked galactose residues." *Biochemistry* 35:3238–3246.

Ueno, S., K. Takeda, F. Izumi, M. Futai, W. Schwarz and M. Kawamura. 1997. "Assembly of the chimeric Na$^+$/K$^+$-ATPase and H$^+$/K$^+$-ATPase β-subunit with the Na+/K+-ATPase α-subunit." *Biochim. Biophys. Acta* 1330:217–224.

Vagin, O., E. Tokhtaeva and G. Sachs. 2006. "The role of the β$_1$-subunit of the Na$^+$,K$^+$-ATPase and its glycosylation in cell-cell adhesion." *J. Biol. Chem.* 281:39573–39587.

Vagin, O., S. Denevich and G. Sachs. 2003. "Plasma membrane delivery of the gastric H$^+$,K$^+$-ATPase: the role of β-subunit glycosylation." *Am. J. Physiol. Cell Physiol.* 285:C968–C976.

Vagin, O., S. Turdikulova and E. Tokhtaeva. 2007. "Polarized membrane distribution of potassium-dependent ion pumps in epithelial cells: Different roles of the N-glycans of their β-subunits." *Cell. Biochem. Biophys.* 47:376–391.

Vagin, O., S. Turdikulova and G. Sachs. 2004. "The H$^+$,K$^+$-ATPase β-subunit as a model to study the role of N-glycosylation in membrane trafficking and apical sorting." *J. Biol. Chem.* 279:39026–39034.

Vagin, O., S. Turdikulova and G. Sachs. 2005a. "Recombinant addition of N-glycosylation sites to the basolateral Na$^+$,K$^+$-ATPase β$_1$-subunit results in its clustering in caveolae and apical sorting in HGT-1 cells." *J. Biol. Chem.* 280:43159–43167.

Vagin, O., S. Turdikulova, I. Yakubov and G. Sachs. 2005b. "Use of the H$^+$,K$^+$-ATPase β-subunit to identify multiple sorting pathways for plasma membrane delivery in polarized cells." *J. Biol. Chem.* 280:14741–14754.

Valenti, G., J. M. Verbavatz, I. Saboli?, D. A. Ausiello, A. S. Verkman and D. Brown. 1994. "A basolateral CHIP28/MIP26-related protein (BLIP) in kidney principal cells and gastric parietal cells." *Am. J. Physiol.* 267:C812–C820.

Vallon, V., F. Grahammer, H. Volkl, C. D. Sandu, K. Richter, R. Rexhepaj, U. Gerlach, Q. Rong, K. Pfeifer and F. Lang. 2005. "KCNQ1-dependent transport in renal and gastrointestinal epithelia." *Proc. Natl. Acad. Sci. U. S. A.* 102:17864–17869.

van der Hijden, H. T., E. Grell, J. J. de Pont and E. Bamberg. 1990. "Demonstration of the electrogenicity of proton translocation during the phosphorylation step in gastric H^+,K^+-ATPase." *J. Membr. Biol.* 114:245–256.

Vasilyev, A., K. Khater and R. F. Rakowski. 2004. "Effect of extracellular pH on presteady-state and steady-state current mediated by the Na^+,K^+ pump." *J. Membr. Biol.* 198:65–76.

Vilsen, B., J. P. Andersen, J. Petersen and P. L. Jorgensen. 1987. "Occlusion of $^{22}Na^+$ and $^{86}Rb^+$+ in membrane-bound and soluble protomeric α/β-units of Na^+,K^+-ATPase." *J. Biol. Chem.* 262:10511–10517.

Wallmark, B., C. Briving, J. Fryklund, K. Munson, R. Jackson, J. Mendlein, E. Rabon and G. Sachs. 1987. "Inhibition of gastric H^+,K^+-ATPase and acid secretion by SCH 28080, a substituted pyridyl(1,2a)imidazole." *J. Biol. Chem.* 262:2077–2084.

Wallmark, B., G. Sachs, S. Mardh and E. Fellenius. 1983. "Inhibition of gastric H^+,K^+-ATPase by the substituted benzimidazole, picoprazole." *Biochim. Biophys. Acta* 728:31–38.

Wallmark, B., H. B. Stewart, E. Rabon, G. Saccomani and G. Sachs. 1980. "The catalytic cycle of gastric H^+, K^+-ATPase." *J. Biol. Chem.* 255:5313–5319.

Wang, C., M. Eufemi, C. Turano and A. Giartosio. 1996. "Influence of the carbohydrate moiety on the stability of glycoproteins." *Biochemistry* 35:7299–307.

Wang, J., J. B. Velotta, A. A. McDonough and R. A. Farley. 2001. "All human Na^+,K^+-ATPase α-subunit isoforms have a similar affinity for cardiac glycosides." *Am. J. Physiol. Cell. Physiol.* 281:C1336–C1343.

Wang, T., N. Courtois-Coutry, G. Giebisch and M. J. Caplan. 1998. "A tyrosine-based signal regulates H^+,K^+-ATPase-mediated potassium reabsorption in the kidney." *Am. J. Physiol.* 275:F818–F826.

Wolosin, J. M. and J. G. Forte. 1984. "Stimulation of oxyntic cell triggers K^+ and Cl^- conductances in apical H^+,K^+-ATPase membrane." *Am. J. Physiol.* 246:C537–C545.

Wright, E. M. 2001. "Renal Na^+-glucose cotransporters." *Am. J. Physiol. Renal. Physiol.* 280:F10–F18.

Wright, E. M., E. Turk, K. Hager, L. Lescale-Matys, B. Hirayama, S. Supplisson and D. D. Loo. 1992. "The Na^+/glucose cotransporter (SGLT1)." *Acta Physiol. Scand. Suppl.* 607:201–207.

Yoda, A. and S. Yoda. 1982. "Interaction between ouabain and the phosphorylated intermediate of Na^+,K^+-ATPase." *Mol. Pharmacol.* 22:700–705.

Yokoyama, T., S. Kaya, K. Abe, K. Taniguchi, T. Katoh, M. Yazawa, Y. Hayashi and S. Mardh. 1999. "Acid-labile ATP and/or ADP/P_i binding to the tetraprotomeric form of Na^+,K^+-ATPase accompanying catalytic phosphorylation-dephosphorylation cycle." *J. Biol. Chem.* 274:31792–31796.

YuA, Ovchinnikov, V. V. Demin, A. N. Barnakov, A. P. Kuzin, A. V. Lunev, N. N. Modyanov and K. N. Dzhandzhugazyan. 1985. "Three-dimensional structure of Na^+, K^+-ATPase revealed by electron microscopy of two-dimensional crystals." *FEBS Lett.* 190:73–76.

Zamofing, D., B. C. Rossier and K. Geering. 1989. "Inhibition of N-glycosylation affects transepithelial Na^+ but not Na^+,K^+-ATPase transport." *Am. J. Physiol.* 256:C958–C966.

Zolotarjova, N., S. M. Periyasamy, W. H. Huang and A. Askari. 1995. "Functional coupling of phosphorylation and nucleotide binding sites in the proteolytic fragments of Na^+,K^+-ATPase." *J. Biol. Chem.* 270:3989–3995.

Die VDM Verlagsservicegesellschaft sucht für wissenschaftliche Verlage abgeschlossene und herausragende

Dissertationen, Habilitationen, Diplomarbeiten, Master Theses, Magisterarbeiten usw.

für die kostenlose Publikation als Fachbuch.

Sie verfügen über eine Arbeit, die hohen inhaltlichen und formalen Ansprüchen genügt, und haben Interesse an einer honorarvergüteten Publikation?

Dann senden Sie bitte erste Informationen über sich und Ihre Arbeit per Email an *info@vdm-vsg.de*.

Sie erhalten kurzfristig unser Feedback!

VDM Verlagsservicegesellschaft mbH
Dudweiler Landstr. 99   Telefon +49 681 3720 174
D - 66123 Saarbrücken   Fax   +49 681 3720 1749
www.vdm-vsg.de

Die VDM Verlagsservicegesellschaft mbH vertritt

Printed by Books on Demand GmbH, Norderstedt / Germany